DJP

Patricia A. Buffler, Ph.D.
Professor of Epidemiology
University of Texas Health Science
Center at Houston
P. O. Box 20186
Houston, Texas 77025

8-4-03

Brain Tumor Therapy

Progress in Experimental Tumor Research

Vol. 28

Series Editor
F. Homburger, Cambridge, Mass.

KARGER

S. Karger · Basel · München · Paris · London · New York · Tokyo · Sydney

Brain Tumor Therapy

Volume Editors
M. L. Rosenblum and *C. B. Wilson,* San Francisco, Calif.

37 figures and 12 tables, 1984

KARGER

S. Karger · Basel · München · Paris · London · New York · Tokyo · Sydney

Progress in Experimental Tumor Research

National Library of Medicine, Cataloging in Publication
Brain tumor therapy
Volume editors, M. L. Rosenblum and C. B. Wilson, – Basel; New York: Karger, 1984.
(Progress in experimental tumor research; v. 28)
Companion volume to: Brain tumor biology.
1. Brain Neoplasms – therapy I. Rosenblum, M. L. (Mark L.) II. Wilson, Charles B., 1929 – III. Title: Brain tumor therapy IV. Series
W1 PY668T v. 28 [WL 358 B8134]
ISBN 3-8055-3699-2

Drug Dosage

The authors and the publisher have exerted every effort to ensure that drug selection and dosage set forth in this text are in accord with current recommendations and practice at the time of publication. However, in view of ongoing research, changes in government regulations, and the constant flow of information relating to drug therapy and drug reactions, the reader is urged to check the package insert for each drug for any change in indications and dosage and for added warnings and precautions. This is particularly important when the recommended agent is a new and/or infrequently employed drug.

Printed in Switzerland by BDV Basler Druck- und Verlagsanstalt, Basel
ISBN 3-8055-3699-2

Contents

Brain Tumor Therapy

The papers with emphasis on 'Brain Tumor Therapy' are published in this volume. The papers with emphasis on 'Brain Tumor Biology' are published in vol. 27 of the series (see table of contents on p. VI).

Contents

Brain Tumor Biology

Published as vol. 27 in the series Progress in 'Experimental Tumor Research'.

Contents

Contributors List

Henry M. Bartkowski, MD, PhD, Assistant Professor of Neurological Surgery, School of Medicine, University of California, San Francisco, and Assistant Chief, Neurosurgery Service, San Francisco General Hospital Medical Center, San Francisco, Calif., USA

Mark Bernstein, MD, Research Fellow, Departments of Neurological Surgery and Radiation Oncology, Brain Tumor Research Center, School of Medicine, University of California, San Francisco, San Francisco, Calif., USA; currently Resident in Neurosurgery, University of Toronto, Toronto, Ont., Canada

Darell D. Bigner, MD, PhD, Professor of Pathology, Division of Neuropathology, Duke University Medical Center, Durham, N.C., USA

Sandra H. Bigner, MD, Associate Professor of Pathology, and Director, Section on Cytogenetics, Division of Cytopathology and Cytogenetics, Duke University Medical Center, Durham, N.C., USA

Ronald G. Blasberg, MD, Senior Investigator, Laboratory of Chemical Pharmacology, Membrane Transport Section, National Cancer Institute, National Institutes of Health, Bethesda, Md., USA

William J. Bodell, PhD, Assistant Research Biochemist, Department of Neurological Surgery, Brain Tumor Research Center, School of Medicine, University of California, San Francisco, San Francisco, Calif., USA

Mario A. Bourdon, PhD, Predoctoral Fellow, Department of Pathology, Duke University Medical Center, Durham, N.C., USA

Richard H. Britt, MD, PhD, Assistant Professor of Neurosurgery, Stanford University School of Medicine, Stanford, Calif., USA

Rodney A. Brooks, PhD, Research Physicist, National Institute of Neurological and Communicative Disorders and Stroke, National Institutes of Health, Bethesda, Md., USA

Christopher E. Cann, PhD, Associate Professor of Radiology, School of Medicine, University of California, San Francisco, San Francisco, Calif., USA

Stefan Carrel, PhD, Unit of Human Cancer Immunology, Lausanne Branch, Ludwig Institute for Cancer Research, Epalinges, Switzerland

Joseph R. Castro, MD, Professor of Radiation Oncology, and Vice Chairman, Department of Radiation Oncology, School of Medicine, University of California, San Francisco, and Director of Radiotherapy, Lawrence Berkeley Laboratory, Berkeley, Calif., USA

R. Edward Coleman, MD, Director of Nuclear Medicine, Department of Radiology, Duke University Medical Center, Durham, N.C., USA

Craig J. Cummins, PhD, Senior Staff Fellow, Surgical Neurology Branch, National Institute of Neurological and Communicative Disorders and Stroke, National Institutes of Health, Bethesda, Md., USA

Richard L. Davis, MD, Professor of Pathology (Neuropathology), Neurological Surgery, and Neurology, School of Medicine, University of California, San Francisco, San Francisco, Calif., USA

Stephen J. DeArmond, MD, PhD, Assistant Professor of Pathology, School of Medicine, University of California, San Francisco, San Francisco, Calif., USA

Dennis F. Deen, PhD, Associate Professor of Neurological Surgery and Radiation Oncology, Brain Tumor Research Center, School of Medicine, University of California, San Francisco, San Francisco, Calif., USA

Leo de Ridder, SM, Senior Scientist, Department of Experimental Cancerology, Radiotherapy and Nuclear Medicine, Academic Hospital, DePintelaan, Ghent, Belgium

Nicolas de Tribolet, MD, Privat-Docent, Neurosurgical Service, Centre Hospitalier Universitaire Vaudois, Lausanne, Switzerland

William C. Dewey, PhD, Professor, Department of Radiation Oncology, School of Medicine, University of California, San Francisco, San Francisco, Calif., USA

Giovanni Di Chiro, MD, Chief, Neuroradiology and Computerized Tomography Section, Surgical Neurology Branch, National Institute of Neurological and Communicative Disorders and Stroke, National Institutes of Health, Bethesda, Md., USA

Bruce G. Douglas, BASc, MD, FRCP (C), Radiation Oncology, Vancouver, B.C., Canada

Lawrence F. Eng, PhD, Professor of Pathology (Research), Stanford University School of Medicine, Stanford, Calif., and Chief, Chemistry Section Laboratory Service, Veterans Administration Medical Center, Palo Alto, Calif., USA

John R. Fike, PhD, Assistant Professor of Radiation Oncology, Radiology and Neurosurgery, School of Medicine, University of California, San Francisco, San Francisco, Calif., USA

Howard J. Fingert, MD, Research Fellow in Medicine, Division of Cell Growth and Regulation, Dana-Farber Cancer Institute and Hematology-Oncology Unit, Massachusetts General Hospital, Harvard Medical School, Boston, Mass., USA

Henry S. Friedman, MD, Assistant Professor, Division of Hematology-Oncology, Department of Pediatrics, Duke University Medical Center, Durham, N.C., USA

Massimo A. Gerosa, MD, Assistant Professor, Istituto di Neurochirurgia, Università di Verona, Italy

G. Yancey Gillespie, PhD, Research Associate Professor of Surgery and Pathology, Division of Neurological Surgery, and Member of the Cancer Research Center, University of North Carolina, Chapel Hill, N.C., USA

Dennis R. Groothuis, MD, Assistant Professor of Neurology, Northwestern University, Evanston Hospital, Evanston, Ill., USA

Philip H. Gutin, MD, Associate Professor of Neurological Surgery and Radiation Oncology, Brain Tumor Research Center, School of Medicine, University of California, San Francisco, San Francisco, Calif., USA

Mary M. Herman, MD, Professor of Pathology, Division of Neuropathology, University of Virginia School of Medicine, Charlottesville, Va., USA

Fred H. Hochberg, MD, Department of Neurology, Harvard Medical School, and Neurology Service, Massachusetts General Hospital, Boston, Mass., USA

Eugene V. Holahan, Jr., PhD, Captain, United States Army, Research Biophysicist, Armed Forces Radiobiology Research Institute, Bethesda, Md., USA

Takao Hoshino, MD, DMSc, Professor of Neurological Surgery, Brain Tumor Research Center, School of Medicine, University of California, San Francisco, San Francisco, Calif., USA

Paul Kleihues, MD, Professor and Director, Division of Neuropathology, Institute of Pathology, University of Freiburg, Freiburg, FRG

Paul L. Kornblith, MD, Chief, Surgical Neurology Branch, National Institute of Neurological and Communicative Disorders and Stroke, National Institutes of Health, Bethesda, Md., USA

Ole Didrik Laerum, MD, Professor of Experimental Pathology and Oncology, The Gade Institute, Department of Pathology, University of Bergen, Haukeland Hospital, Norway

A. A. Lammertsma, MRC Cyclotron Unit, Hammersmith Hospital, London, England

Bernard E. Lyons, BA, Research Associate, Division of Neurosurgery, Department of Surgery, Stanford University School of Medicine, Stanford, Calif., USA

Jean-Pierre Mach, Associate Professor, Unit of Human Cancer Immunology, Lausanne Branch, Ludwig Institute for Cancer Research, Epalinges, Switzerland

M. S. Mahaley, Jr., MD, PhD, Professor of Neurological Surgery and Neurobiology, University of North Carolina School of Medicine, Chapel Hill, N.C., USA

Joachim Mark, MD, Associate Professor of Pathology, and Head, Laboratory of Cytogenetics, Central Hospital, Skövde, Sweden

Laurence J. Marton, MD, Professor of Laboratory Medicine and Neurological Surgery, Brain Tumor Research Center, and Chairman, Department of Laboratory Medicine, School of Medicine, University of California, San Francisco, San Francisco, Calif., USA

Peter Molnar, MD, Fogarty Research Fellow, Laboratory of Chemical Pharmacology, Membrane Transport Section, National Cancer Institute, National Institutes of Health, Bethesda, Md., USA; currently with the Department of Pathology, University Medical School of Debrecen, Debrecen, Hungary

Sverre J. Mørk, MD, Consultant in Neuropathology, The Gade Institute, Department of Pathology, University of Bergen, Haukeland Hospital, Norway

Edward A. Neuwelt, MD, Associate Professor of Neurosurgery, Assistant Professor of Biochemistry, School of Medicine, Oregon Health Sciences University, Portland, Oreg., USA

Stina M. Oredsson, MS, Visiting Predoctoral Fellow, Brain Tumor Research Center, Department of Neurological Surgery, School of Medicine, University of California, San Francisco, San Francisco, Calif., USA; currently Doctoral Candidate, Department of Zoophysiology, University of Lund, Sweden

Nicholas J. Patronas, MD, Staff Radiologist, National Institute of Neurological and Communicative Disorders and Stroke, National Institutes of Health, Bethesda, Md., USA

Douglas W. Pounds, BS, Research and Development Engineer/Physicist, Division of Neurosurgery, Department of Surgery, Stanford University School of Medicine, Stanford, Calif., USA

Manfred F. Rajewsky, MD, Professor and Director, Institute of Cell Biology (Cancer Research), University of Essen, Essen, FRG

C. G. Rhodes, MRC Cyclotron Unit, Hammersmith Hospital, London, England

Mark L. Rosenblum, MD, Associate Professor of Neurological Surgery, Brain Tumor Research Center, School of Medicine, University of California, San Francisco, San Francisco, Calif., USA

Lucien J. Rubinstein, MD, Professor of Pathology, and Director, Division of Neuropathology, University of Virginia School of Medicine, Charlottesville, Va., USA

Michael Salcman, MD, FACS, Associate Professor of Neurological Surgery, and Chief, Neuro-Oncology Service, University of Maryland School of Medicine and Hospital, Baltimore, Md., USA

S. Clifford Schold, Jr., MD, Associate Professor of Neurology, Department of Medicine, Duke University Medical Center, Durham, N.C., USA

Joan Rankin Shapiro, PhD, Research Associate, Department of Neurology and George C. Cotzias Laboratory of Neuro-Oncology, Memorial Sloan-Kettering Cancer Center, Cornell University Medical College, New York, N.Y., USA

William R. Shapiro, MD, Professor of Neurology, and Head, George C. Cotzias Laboratory of Neuro-Oncology, Memorial Sloan-Kettering Cancer Center, Cornell University Medical College, New York, N.Y., USA

Glenn E. Sheline, MD, PhD, Professor of Radiation Oncology, and Vice Chairman, Department of Radiation Oncology, School of Medicine, University of California, San Francisco, San Francisco, Calif., USA

Barry H. Smith, MD, PhD, Deputy Chief, Surgical Neurology Branch, National Institute of Neurological and Communicative Disorders and Stroke, National Institutes of Health, Bethesda, Md., USA; currently with the Neurosurgical Service, Memorial Sloan-Kettering Cancer Center, New York, N.Y., USA

David J. Stewart, MD, FRCP (C), Clinical Assistant Professor of Medicine, University of Ottawa, Medical Oncologist, Ontario Cancer Foundation, Ottawa Clinic, General Hospital Division, Ottawa, Ont., Canada

Ronald E. Talcott, PhD, Associate Professor of Laboratory Medicine, Brain Tumor Research Center, School of Medicine, University of California, San Francisco, San Francisco, Calif., USA

David G. T. Thomas, MA, MRCP, FRCS, Ed., Consultant Neurosurgeon and Senior Lecturer, Gough-Cooper Department of Neurological Surgery, The National Hospital, Queen Square, London, England

Philip J. Tofilon, PhD, Assistant Research Pharmacologist, Departments of Neurological Surgery and Radiation Oncology, Brain Tumor Research Center, School of Medicine, University of California, San Francisco, San Francisco, Calif., USA

Scott R. VandenBerg, MD, PhD, Resident in Pathology, Division of Neuropathology, University of Virginia School of Medicine, Charlottesville, Va., USA

Charles B. Wilson, MD, Professor of Neurological Surgery, Chairman, Department of Neurological Surgery, and Director, Brain Tumor Research Center, School of Medicine, University of California, San Francisco, San Francisco, Calif., USA

Richard J. S. Wise, MD, MRC Cyclotron Unit, Hammersmith Hospital, London, England

Preface

When Dr. *Thomas Karger* suggested the need for a companion volume on brain tumor therapy to accompany the monograph on *Brain Tumor Biology* in this series on 'Progress in Experimental Tumor Research', we recognized the timeliness of his idea. We saw the value of gathering into a single reference source the experience of experts in the various areas of the field, many of which have never been combined in such a book.

This volume focuses on innovative therapies conceived and directed on the basis of the recent laboratory research findings. Its 15 chapters represent work that is being conducted in institutions actively engaged in brain tumor therapy. The authors review the areas of specificity of treatment, cytotoxic and noncytotoxic treatments, and local therapy. They describe new means of administering cytotoxic agents to improve the therapeutic index, discuss biological response modifiers, and review the status of local therapy with irradiation and hyperthermia. Rather than merely summarizing work presented in previously published material, they include their speculations on the reasons for the success and failure of the individual therapeutic modality and its place among other methods.

In compiling these chapters, we have asked the authors to review concisely the well-established data in their area and the completed results and published concepts underlying their own work. However, the primary emphasis in each chapter is the relevant preliminary findings in a particular area of interest that will serve as the basis for research over the next several years. We believe that this book and its companion volume will serve as a guide for future investigations and possibly new directions for brain tumor therapy.

Our selection of the topics presented reflects personal bias, but selectivity was required because of necessary limitations on the volume's size. Several areas not represented here in which there recently has been significant progress are included in an appendix of selected readings, compiled with the aid of our authors to provide pertinent references that will supplement the work represented in the invited chapters. We hope that the reader interested in topics that are not covered will understand the restrictions of this or any other monograph concerned with evolving and rapidly changing clinical research.

M. L. Rosenblum and *C. B. Wilson*

Acknowledgment

The timely editing of this monograph could not have been accomplished without the superb collaboration of the editorial staff of our Department of Neurological Surgery at the University of California, San Francisco. For this, we are indebted to *Susan Eastwood-Berry,* and *Beverly H. McGehee.*

To our patients

Specificity of Treatment

Prog. exp. Tumor Res., vol. 28, pp. 1–17 (Karger, Basel 1984)

Stem Cell Sensitivity[1]

M. L. Rosenblum[2], *M. A. Gerosa*

Brain Tumor Research Center of the Department of Neurological Surgery, School of Medicine, University of California, San Francisco, Calif., USA

Introduction

Malignant tumors are composed of tumor and normal host cells such as fibroblasts, endothelial cells, macrophages and lymphocytes that are held within a connective tissue matrix. An alteration in any component will cause the tumor to change size in situ. Tumor cells themselves exist in various states of intactness and have different proliferative capacities that depend on local tumor factors such as the degree of oxygenation and nutritional status, and on inherent cellular determinants. Furthermore, a tumor cell that is 'killed' by a drug or radiation may divide several times before it dies [3]. Finally, removal of the dead cells from the tumor mass is delayed [12]. Because of this complexity, it is not surprising that studies of changes in the size of a tumor during growth and after treatment have contributed disappointingly little to our ability to develop more effective treatment methods for patients harboring malignancies.

Recently, emphasis has been focused on clonogenic tumor cells that have the potential for unlimited proliferation. Continuous growth of a malignant tumor reflects multiplication of these cells; and a therapy that could destroy all clonogenic cells should produce a cure. It is assumed that a better understanding of both tumor biology and the effectiveness of treatment will depend on knowledge about this clonogenic population. The number and types of clonogenic cells within a solid tumor, their

1 Supported in part by National Cancer Institute Grants No. CA-13525, CA-28519, and CA-31882.

2 M.L.R. is recipient of Teacher Investigator Development Award No. 1K-07NS 00604 NSPA from NINCDS.

growth kinetics, and their differential sensitivity to cytotoxic agents probably account for variations in tumor growth and responsiveness that have been found among tumors with similar histologic characteristics.

In vitro colony formation assays have gained repute as the method of choice for analysis of clonogenic cell survival. This approach has been used to better understand tumor growth and the effects of drugs and irradiation on many established animal and human tumor cell lines.

The clonal growth of cells derived from human tumors was first described by *Hamburger and Salmon* [7] who used a 2-layered, soft agar technique to grow multiple myeloma and ascitic ovarian carcinoma cells in vitro. *Courtenay and Selby* [2] grew colonies from single cells derived from solid tumors of various types both in a soft agar culture system with a replenishable liquid phase, added red blood cells, and low O_2 concentration, and with an agar diffusion-chamber method. Colonies were obtained in their studies from 31 of 48 tumors with colony forming efficiencies (CFE) of between 0.01 and 15% (average = 2%); results were similar for both methods. *Wells* et al. [36] and *Rosenblum* et al. [26] described clonogenic growth of cells from a variety of tumors and brain tumors, respectively, using monolayer techniques and liquid medium. CFEs for cells using *Courtenay and Selby's* [2] method appear to be higher than CFEs of cells grown with *Hamburger and Salmon's* [7] soft agar or monolayer [26, 36] techniques. Results of 2 recent studies have suggested that colonies grow more readily using a low O_2 method such as *Courtenay's* compared to soft agar systems such as *Hamburger's* [8, 32b]. Furthermore, cell survival studies have given different results when they were conducted using these 2 different techniques [8, 32]. However, a definitive comparison between the methods as a general statement of applicability is not possible at this time.

The potential for in vitro tumor cell sensitivity testing as a predictive test for in situ patient response to chemotherapeutic agents was demonstrated first by *Salmon* et al. [28]. They found a correlation between in vitro and in situ sensitivity for 31 of 32 drug trials in 18 patients harboring multiple myeloma and ovarian carcinoma. Similar results have been reported subsequently [19, 20, 27, 33]. The ability of this assay to predict tumor resistance to chemotherapy is 85–97%, whereas the prediction of sensitivity is 45–70% for heavily pretreated patients [27, 33]. These figures are similar to the predictive efficiency of the estrogen-receptor assay, a method considered useful in planning therapy for patients with breast tumors [18].

Commenting on the work of *Salmon* et al. [28], *Frei and Lazarus* [6] summarized the potential implications of studies of tumor cell sensitivity:

'This assay, based on current understanding of tumor biology, represents a promising approach to the in vitro assessment of tumor behavior and response to therapy. The highly positive correlation between the test results and in vivo human tumor response observed by *Salmon* et al. most certainly supports this approach. Continuing advances in tumor biology generally, and cell culture particularly, can be expected to solve many of the problems, with resultant further improvement and broader application of the technique. An effective and practical predictive test for antitumor agents would have profound effect on the treatment of cancer.'

Methodological Considerations

The in vitro sensitivity of cultured cells to chemotherapeutic agents depends upon several factors:

(1) Cellular exposure dose is determined by the duration of cellular exposure to the active drug moiety. The amount of active drug available depends upon the mode and rate of drug activation, binding, and breakdown, all of which may be modified by medium constituents; for example, 20% fetal calf serum increases the rate of BCNU breakdown and decreases the in vitro cell exposure dose [38].

(2) Cytotoxicity depends on the inherent sensitivity of tumor cells to a drug and on the mechanism of drug activity when the state of cell proliferation plays a role. For example, cytosine arabinoside (Ara-C), a cell-cycle, phase-specific agent, kills only cells that are in S phase [30]. Cell kill should be greater for cells rapidly growing in long-term monolayer passage, which have a large growth fraction and LI, than for cells in slowly growing tumors in situ or in low growth fraction primary culture passage (low LI) [9]. In addition, cell kill can be maximum only if cells are exposed to drug for the length of a cell cycle, during which all cells will pass through S phase [30]. Therefore, the effect of short-term (1–2 h) and long-term drug dosage will be different for cell-cycle phase-specific agents.

(3) Finally, cells may repair potentially lethal damage (PLD) caused by drug if the cells are left undisturbed in a nonproliferative state for a prolonged period before plating for the CFE assay [1]. This phenomenon more closely parallels the occurrences in situ than immediate replating for assay with the attendant trauma of enzyme treatment and stimulation of proliferation. This factor must be considered when analyzing in vitro drug treatments.

To accurately relate tumor cell sensitivity in vitro to the potential sensitivity of the same tumor cells in vivo, one must have a reasonable expectation of the in vivo drug dose achievable in the tumor when a patient is treated at or near the maximally tolerated dose (MTD). This concept has not been given important consideration in the past because it is difficult to measure or estimate drug exposure of tumor cells in situ [15, 17, 37].

To determine the appropriate dose level and dose exposure period for in vitro cell survival studies, much must be known about a drug's mode of action, its dependence on duration of exposure, and a multitude of pharmacokinetic factors. For drugs such as the

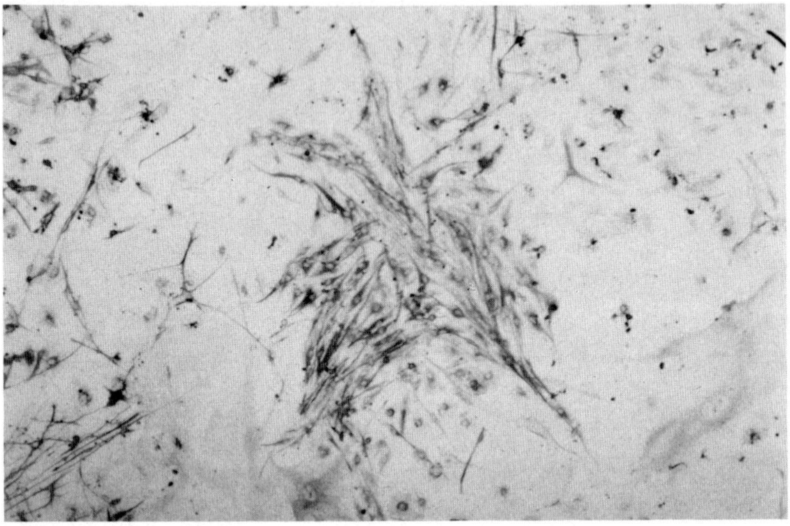

Fig. 1. Colony containing cells some of which demonstrate the intracellular presence of glial fibrillary acidic protein. Horseradish peroxide stain with rabbit-GFAP antibody kindly supplied by *L. Eng.* ×66.

chloroethylnitrosoureas (CENU), the total drug exposure integral (C_t) determines the amount of cell kill [17, 34]. For most prodrug alkylating agents, which are activated by linear reactions, dependence on C_t would also be expected. Agents that require a threshold period of drug exposure to initiate cell kill, such as dianhydrogalactitol, would be an exception to that rule [16].

Knowledge of the approximate in vivo tumor cell drug exposure, determined from plasma pharmacokinetic data [13, 15], by direct measurements of drug concentrations within tumors [31], or by mathematical modeling methods [15], must be augmented by an understanding of the actual in vitro cell exposure. Factors that will influence the in vitro treatment include drug stability, transformation to more or less active drug moieties in the medium, cellular permeability, and intracellular biotransformation [15, 37].

Development of the Brain Tumor Cell Assay

Recently, we developed an in vitro clonogenic cell assay for the investigation of human brain tumor cells [26]. The procedure for obtaining monolayer clonal outgrowths of cells derived from human brain tumor specimens has been described [26]. Briefly, it entails obtaining solid tumor biopsies from the operating room and mincing them into pieces less than 1 mm^3, which are treated for 30 min with an enzyme cocktail consisting of pronase, collagenase, and DNAse. This disaggregation method produces the greatest clonogenic cell yield and the purest suspension of single cells from biopsy specimens among a number of dissociation techniques tested [20]. After filtration to remove nondigested particles, the

Table I. Monoclonal antibody recognition

Tumor No.	Culture diagnosis	Passage No.	Hybridoma antibody-binding ratio		
			GE 2	BF 7	CG12
126	GM	51	4	2	2
159	GM	9	6	3	3
167	GM	14	14	9	6
168	GM	10	14	8	6
188	GM	5	22	15	5
253	GM	6	4	3	2
259	GM	5	3	2	2
210	AA	9	4	3	3
160	AB	10	14	9	4
181	AB	4	4	10	2
213	MB	8	3	2	1.5
242	EP	7	1	1	1
254	EP	7	2	1.5	2
185	NL	3	3	2	2
Total positive/number tested = T			12/14	8/14	5/14

GM = Glioblastoma multiforme; AA = anaplastic astrocytoma; AB = astroblastoma (same patient, 2 surgeries); MB = medulloblastoma; EP = ependymoma; NL = normal brain.

single cell suspension is aliquoted into 60-mm plastic Petri dishes at various dilutions. Plates are incubated for 2–4 weeks at 37°C in a humidified atmosphere with 5% CO_2, in an enriched medium containing 20–30% fetal calf serum and heavily irradiated 9L rat brain tumor feeder cells. A collection of more than 25 cells of similar morphology was considered to be a colony that had originated from a single 'clonogenic' tumor cell derived from the initial biopsy specimen. Colonies have been grown from the majority of tumors analyzed, including malignant gliomas, astrocytomas, oligodendrogliomas, medulloblastomas, ependymomas, meningiomas, choroid plexus carcinomas, and metastatic brain tumors.

A peroxidase-antiperoxidase (PAP) technique for immunohistochemical staining of glial fibrillary acidic protein (GFAP) described by *Eng and Rubinstein* [4] was modified for use on our cultured cells. Preliminary studies show positive results in 4 of 7 malignant gliomas analyzed in early (less than 10) culture passage (fig. 1).

Identification of glioma-associated antigens will facilitate characterization of cells and colonies derived from human brain tumors. A technique for producing somatic cell hybrids that secrete monoclonal antibodies against single antigens was first developed by *Kohler and Milstein* [11]. Hybridomas that are grown and which produce monoclonal antibodies should provide a powerful research tool for the identification of cancer cells. *Schnegg* et al. [29a]

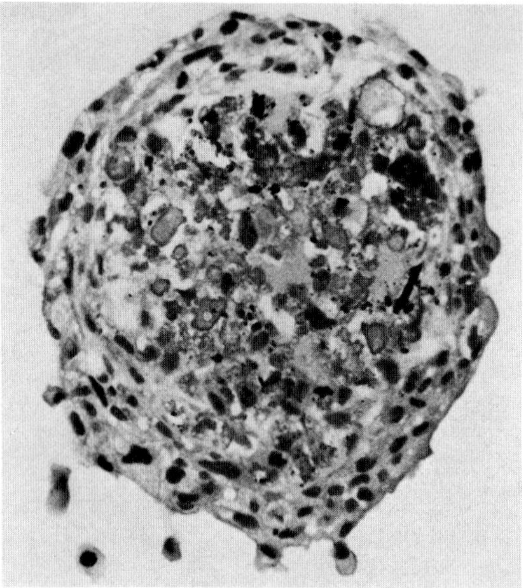

Fig. 2. Three-dimensional multicellular spheroid developed from cells derived from early passage culture of a glioblastoma culture No. 209. HE. ×43.

developed three hybrids that secrete antibodies against malignant glioma cell antigens; the specificity of those monoclonal antibodies are discussed in *de Tribolet* et al. [32a]. They have reacted several of our cell cultures against hybridoma-produced monoclonal antibodies GE 2, BF 7 and CG 12; antigen recognition was determined by a radioimmunoassay [29] (table I). The results are expressed as a binding ratio representing the cell bound counts using the hybridoma antibodies divided by the counts using the control fluid. A ratio of 3 or more is considered significantly positive and implies antigen recognition by the monoclonal antibody. 11 of the 13 tumors tested were recognized by GE 2, 8 of 13 by BF 7, and 5 of 13 by CG 12. The negative results using GE 2 were obtained with 2 ependymomas (HU 242 and 254). Cells proliferating in cultures of normal brain were recognized by GE 2.

Multicellular spheroids have been grown from human brain tumor cells aggregated from early culture passage [25]. These spheroids grow in culture and have histologic appearances that suggest the cells are malignant (fig. 2). Finally, preliminary experiments have shown that transplantation of cells harvested from colonies grew into tumors in immunodeficient mice [26] and under the renal capsule in irradiated rats [*Weizsaecker,* unpublished observations]. Therefore, we are confident that the results of GFAP staining, monoclonal antibody recognition, the ability to form spheroids, and the growth of tumors in nude mice and under the renal capsule of rats indicate that the colonies represent the growth of clonogenic malignant glial tumor cells.

Preliminary Results

To date, biopsy specimens from 71 brain tumors, including 30 glioblastomas, 22 anaplastic astrocytomas, 7 other malignant gliomas, and 12 other brain tumors, have been analyzed successfully for BCNU sensitivity. Thiotepa, cis-platinum, dianhydrogalactitol, elliptinium, VM-26, and AZQ have been studied in cells obtained from a few patients; results of these studies will not be discussed here. Cell survival analysis demonstrated a maximum of 99% cell kill within the clinically achievable dose range of BCNU. In our initial protocol, cells obtained from a biopsy specimen were treated with BCNU in MEM for 2 h with continuous stirring in spinner flasks and subsequent plating for colony formation. Because BCNU dose-response curves were similar for cells treated immediately after dissociation from tumor biopsy specimens and for cells tested after a few passages in culture [35], subsequent experiments were performed on tumor cells in early passage culture. A 2-hour treatment with 2.5 μg/ml results in a cell exposure dose of 12 $\mu M \cdot h$ which is the approximate maximal clinically achievable dose in the tumor [*Levin*, personal commun.]. The expression of BCNU exposure dose in $\mu M \cdot h$ (C_t) supercedes the expression of dose calculated as the active product formed in culture medium according to *Weinkam and Deen* [31]; 12 $\mu M \cdot h$ is equivalent to the BCNU dose of 8.5 μM found in our earlier publications [20–25]. At that dose and treatment duration a cell kill of 40% discriminates between tumor response and progression in situ.

In our current protocol, early passage tumor cells are treated with BCNU in complete medium (MEM with 20% fetal calf serum) without subsequent medium change. Because the presence of serum increases the rate of extracellular breakdown of BCNU [34], this continuous treatment technique requires a higher BCNU concentration to achieve a cell exposure equivalent to that used in earlier protocols. We determined that a concentration of 5 μg/ml results in a cell exposure of 12 $\mu M \cdot h$ and that a cell kill of 80% discriminates between sensitivity and resistance in vitro at that dose when the continuous exposure technique is used. The reason for the different cell kill discrimination levels is being investigated presently.

We have attempted to analyze cells obtained from tumor biopsy specimens from 79 patients; 90% of the studies were successfully performed (table II). Cells from 34 of 71 specimens (48%) were sensitive to BCNU in vitro. 25 patients, whose cells were tested with BCNU in the assay, were treated with BCNU or another CENU either before or after

Table II. Tumors studied for BCNU sensitivity in vitro

Diagnosis	Culture Results		
	sensitive	resistant	not evaluable[1]
Glioblastoma	13	17	1
Anaplastic astrocytoma	13	9	3
Other malignant gliomas	3	4	1
Astrocytoma	1	1	1
Oligodendroglioma	1	1	–
Ependymoma	2	1	2
Medulloblastoma	1	1	–
Chordoma	0	1	–
Metastatic lung tumor	0	2	–
Total (79)	34	37	8

[1]Contaminated, poor growth, or technical problems.

biopsy; the clinical activity of nitrosoureas could be estimated in these patients. Tumor responses were considered to be equivalent for BCNU, CCNU, and PCNU [(1(2-chloroethyl)-3-(2,6-dioxo-3 piperidyl)-1-nitrosourea)] because their activity in clinical trials against gliomas has been shown to be equivalent [5], their mechanism of action is probably the same [34], and they are cross-resistant in vitro [21]. All patients received optimum standard care. Antitumor activity was assessed according to UCSF criteria [14], and the results were evaluated without knowledge of the assay results.

The correlation between sensitivity in culture and patient response was better for clinical resistance than for sensitivity. 5 of 11 tumors (45%) with cells that were sensitive in culture, responded clinically; while all 13 tumors containing cells that were resistant in culture were resistant clinically. The predictive value of our assay can be best evaluated for the 15 patients who had not been treated with nitrosoureas before biopsy specimens were obtained. The assay predicted clinical response for 5 of 7 patients (71%) and clinical resistance for 8/8 patients (100%).

Dose-response curves for the tumor cells obtained from the first 13 patients in the prospective treatment group are shown in figure 3. The separation between the sensitive and resistant tumors is progressively more distinct at BCNU doses higher than the MCAD. Dose-response

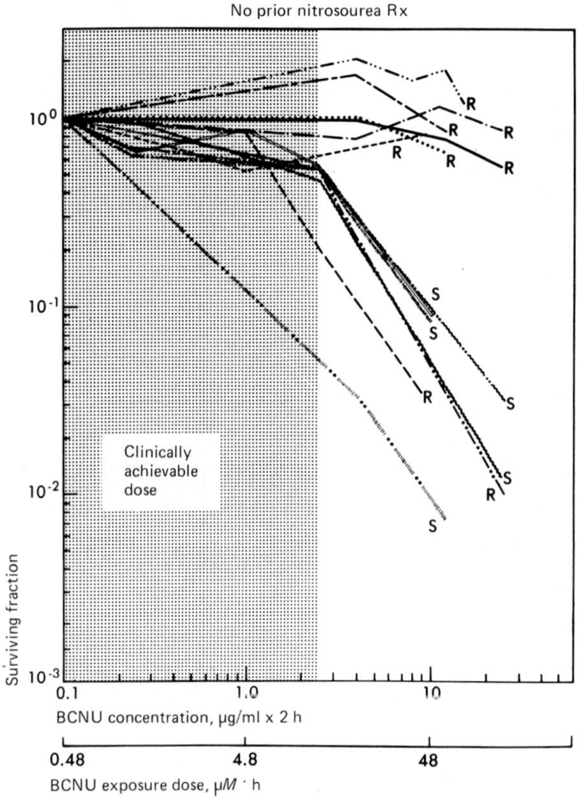

Fig. 3. BCNU dose-response curves for tumor cells obtained from patients who were not treated with nitrosoureas before biopsy. The BCNU dose is represented both as the concentration used over a 2-hour treatment and as the cell exposure dose expressed as $\mu M \cdot h$; the shaded region is the clinically achievable dose range. Each curve represents an individual patient; the letter adjacent to each curve signifies the clinical response to nitrosourea therapy (R = resistant; S = sensitive).

curves for 6 of the 7 sensitive tumors had a shoulder followed by a region of exponential cell kill; the exponential portion of the curve was most obvious at BCNU doses higher than the MCAD. Figure 4 shows dose-response curves for the first 8 patients we evaluated who had failed nitrosourea therapy before cells were obtained for the assay. Cells from 4 tumors were judged to be sensitive at the MCAD: the dose-response curve for one of these tumors had a plateau region at a drug dose that was

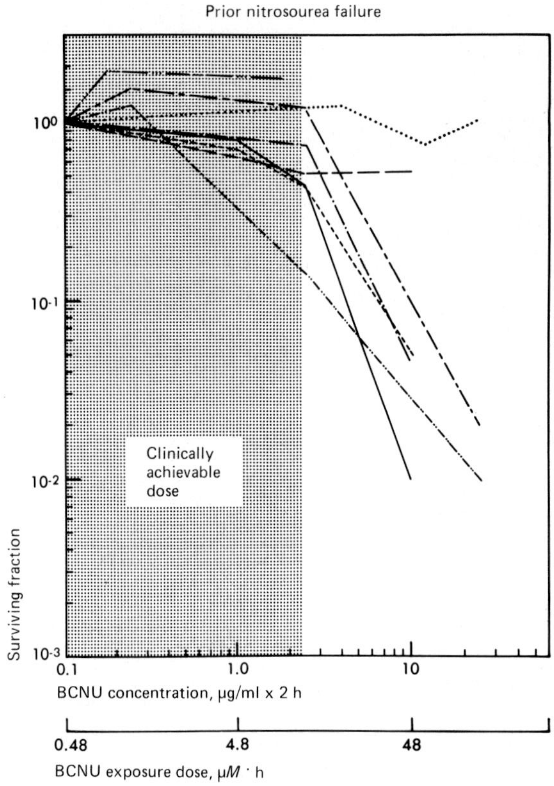

Fig. 4. BCNU dose-response curves for tumor cells derived from patients who were treated with and failed therapy with nitrosoureas before reoperation and in vitro analysis was performed. The BCNU dose is represented both as the concentration used over a 2-hour treatment and as the cell exposure dose expressed as $\mu M \cdot h$; the shaded region is the clinically achievable dose range. Each curve represents an individual patient.

higher than the MCAD. Cells from 4 tumors were resistant to BCNU at the MCAD. 3 of the resistant cell populations were treated with higher doses; dose-response curves for cells from 2 of these tumors had broad shoulders and a region of exponential cell kill only at the highest drug dose used.

Clonogenic cell assays may provide information on the reasons for the failure of patient treatment. Cell survival data for cells treated with up to 10 times the clinically achievable BCNU dose (fig. 3, 4) showed 2 distinct patterns of response: (1) there was essentially no cell kill (8 patients), or (2) there was little or no cell kill in cells from 12 patients at clinically achievable doses, but there was a marked increase in cytotoxicity at higher doses. These findings imply that failure to respond to chemotherapy is the result of inherent tumor cell resistance to drugs in some patients and is the result of inadequate drug delivery in others.

2 patients who responded to BCNU and whose cells were sensitive in the assay underwent a second operation when their tumors recurred and clinical resistance to BCNU had been documented. The recurrent tumors were partially necrotic, and cells obtained at the second procedure showed persistent cell sensitivity to BCNU. This observation suggests that, in some patients, recurrence may reflect insufficient drug delivery to tumor rather than the development of drug resistance [22].

Glioblastomas are among the group of tumors that are the most resistant to therapeutic manipulations. A comparison of the in vitro activity of BCNU against brain tumors [20], ovarian carcinomas, and multiple myeloma [28] confirmed that clonogenic glioblastoma cells are more resistant to BCNU than cells from other human tumors (fig. 5). Analysis of clonogenic cells from malignant brain tumors should permit studies of the cause(s) for tumor resistance to chemotherapeutic agents and may aid in the rational selection of drugs that may have therapeutic efficacy.

We used the clonogenic cell assay to determine if the clinical observation that younger patients with malignant gliomas survive longer than older patients was the result of an inherent, age-related sensitivity of clonogenic cells. A strong inverse relationship was found between in vitro cell kill and patient age [24]. The survival of 9 young patients (less than 50 years of age) was significantly longer (median = 54+ weeks) than for 9 older patients (greater than 50 years of age, median = 36 weeks). All patients had received radiation therapy postoperatively and $\frac{8}{9}$ in each group had been treated with nitrosoureas. The longer survival for the younger group could not be attributed to tumor pathology, size, or location, nor to the Karnofsky score or specific treatment. However, results of the clonogenic cell assay showed that cells for 89% of the younger patients were sensitive to BCNU, while cells from only 11% of older patients were sensitive. Furthermore, patients harboring sensitive tumor cells were significantly younger ($p < 0.005$) (mean = 33 years; range = 9–54) than patients harboring resistant cells (mean = 60 years; range = 36–73). Therefore, age-dependent survival of patients with malignant gliomas is probably the result of an inherent difference in clonogenic tumor cells.

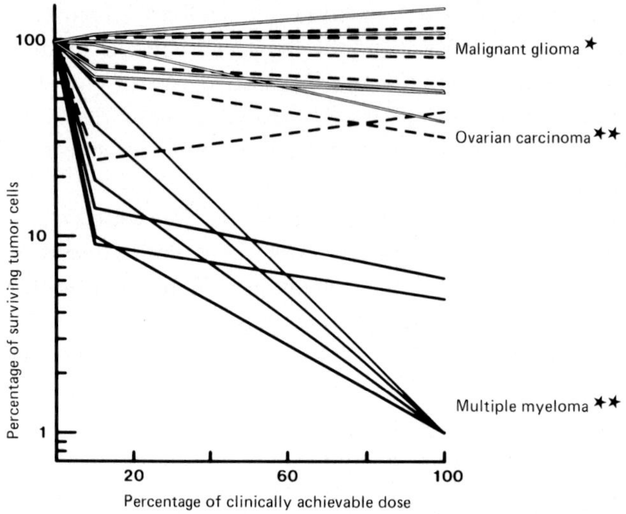

Fig. 5. Comparison of the sensitivity of glioblastoma, ovarian carcinoma and multiple myeloma cells to BCNU treatment in vitro. Dose is expressed according to the clinically achievable dose range (equivalent to 12 $\mu M \cdot$h, *Levin,* personal commun.) Each line represents an individual tumor treated immediately after dissociation from the biopsy specimen; tumors were randomly selected from among experiments performed by *Rosenblum and Dougherty* [22] for malignant gliomas (*; double density light line) and *Salmon* et al. [28] (**; dashed lines = ovarian carcinoma, and solid line = multiple myeloma). [Results kindly supplied from data referred to, but not presented in reference No. 28.]

Tumor cell resistance in older patients may represent age-related changes in cell membranes and/or enzymes that would alter drug transport and metabolism, or may represent more efficient repair of drug-induced damage. Modifications in therapy based on the differences between sensitive and resistant cells might lead to improved treatment of patients with malignant gliomas.

Limitations of the Assay and Future Considerations

The development of tumor disaggregation methods that yield maximum numbers of single, viable, clonogenic cells, the characterization of cells by histochemical methods, and the correlation between in vitro and in vivo results (for both prediction of patient response to BCNU and for the age-associated patient survival) validate our methods for the isolation and evaluation of clinically relevant clonogenic cells from human malignant brain tumors. However, tumor heterogeneity, tumor cell kinetics and in vitro growth conditions may seriously limit methods used to study human tumor stem cells.

Preliminary investigations on the influence of tumor heterogeneity have been performed by studying the chemosensitivity of colonies derived from an individual tumor (possibly an indication of clonal heterogeneity) [35]. Low concentrations of early passage cells from 2 glioblastomas (No. 126 and 159) were plated into 60-mm Petri dishes. Plates were scanned for the purity of the single cell suspension. After 2–4 weeks, individual colonies were scraped from the dishes, transferred to small multiwell plates, and cultured to obtain large numbers of cells. The sensitivity to BCNU of all cultures derived from colonies was compared to the original cell suspensions and an uncloned early culture passage. One cell culture ('Clone No. 2' of 4 derived from tumor No. 159) showed a different dose response to BCNU than the other clones and the parent cell populations from tumor No. 159; cell sensitivity from the clone and parent cells were similar for tumor No. 126 [35]. Of course, these studies must be repeated and expanded before a relevant interpretation on the influence of clonal heterogeneity is made. *Shapiro and Shapiro* [29b] have discussed this issue in detail.

Frequently tumors have histologically distinct regions, and analysis of cells from a given region might not represent all cells in the tumor. To determine whether a biopsy specimen obtained from one area in a large heterogeneous tumor will give results representative of the whole tumor, 2 or 3 grossly distinct tumor regions with different histologic characteristics were selected, passaged separately in culture and analyzed for BCNU sensitivity [23]. The BCNU dose-response curves from early passage cultures were remarkably similar for 6 to 8 tumors studied to date. In only 1 instance did we find a distinct difference in BCNU sensitivity for all cell populations within a single tumor. It is important to note that some laboratories have based their interpretation of a tumor's sensitivity solely on the results of treatment with a single drug dose [33]. If our studies were analyzed for tumor sensitivity in that manner, we would have determined that drug activity on cells from separate specimens was different in ⅓ of cases. It seems reasonable that a method that utilizes several drug doses and incorporates the errors inherent in the techniques would be a more accurate means for determining cell sensitivity for a patient's tumor.

Differences between the in vivo and in vitro environments might influence the activity of chemotherapeutic agents. As an example, differences in tumor cell kinetics will affect the action of agents that have any cell cycle specificity and must be accounted for. This requirement is especially pertinent to agents that are administered clinically by prolonged infusions. In order to observe the maximum antitumor activity in vitro, the length of drug exposure for such agents should be at least equivalent to the cell cycle time of clonogenic cells in culture. Therefore, cell kinetic data are necessary for an interpretation of in vitro cell survival results. Preliminary results from our laboratory [*Pertuiset, Rosenblum,* unpubl. observations] suggest that the in vitro cell cycle time for low passage cells obtained from malignant gliomas is approximately the same as the cell cycle time estimated using in situ stathemokinetic methods [9, 10], but the fraction of cells actively replicating their DNA is markedly larger than the growth fraction of tumor cells in situ. Cell sensitivity studies must take these similarities and differences into account.

In order to accurately investigate tumor stem cells, in vitro culture conditions must be optimized separately for different tumor types. Comparing our monolayer system and a widely used semisolid agar technique, we have analyzed the untreated CFE and BCNU sensitivity of glioma cells from 5 early passage cultures. Although the BCNU-induced cytotoxicity was similar for cells grown with the 2 methods, the unperturbed clonogenic capacity (CFE) of cells from all cultures was significantly higher in the monolayer than for the agar method [unpubl. observations].

We anticipate that future investigations on clonogenic cells will lead to a better understanding of the factors influencing the growth of human brain tumors and the sensitivity of their cells to therapeutic manipulations. Such analyses should provide insights into the reason(s) for therapeutic failure, and may permit the screening of new, potentially useful agents for brain tumor therapy in general, and for individual patients in particular. Furthermore, the methods developed and the results anticipated should be applicable to the investigation of any solid human malignancy.

Acknowledgements

We thank *Dolores V. Dougherty* for her excellent aid over the past 8 years; without her assistance this work could not have been performed. We thank *Ann Schlee* for typing the manuscript, *Beverly McGehee* for preparation, and *Neil Buckley* for editing it.

References

1 Bertrand, M.; Deen, D.F.: Factors influencing the recovery from potentially lethal damage (PLD) in mammalian cells in vitro and in vivo. Cancer Treat. Ref. *7:* 1–15 (1980).

2 Courtenay, V.D.; Selby, P.J.; Smith, I.F.; Mills, J.; Peckam, M.J.: Growth of human tumor cell colonies from biopsies using two soft-agar techniques. Br. J. Cancer *38:* 77–81 (1978).

3 Ehmann, U.K.; Wheeler, K.T.: Cinemicrographic determination of cell progression and division abnormalities after treatment with 1,3-bis(2-chloroethyl)-1-nitrosourea. Eur. J. Cancer *15:* 461–473 (1979).

4 Eng, L.F.; Rubinstein, L.J.: Contributions of immunohistochemistry to diagnostic problems of human cerebral tumors. J. Histochem. Cytochem. *26:* 513–522 (1978).

5 Fewer, D.; Wilson, C.B.; Boldrey, E.B.; Enot, K.J.: Phase II study of 1-(2-chloroethyl)-3-cyclohexyl-1-nitrosourea (CCNU; NSC-79037) in the treatment of brain tumors. Cancer Chemother. Rep. *56:* 421–427 (1972).

6 Frei, E., Lazarus, H.: Predictive tests for cancer chemotherapy (Editorial). New Engl. J. Med. *298:* 1358–1359 (1978).

7 Hamburger, A.W.; Salmon, S.E.: Primary bioassay of human tumor stem cells. Science *197:* 461–463 (1977).

8 Hill, B.T.; Rupniak, H.T.; Whelan, R.D.H.; Metcalfe, S.A.: Improved colony formation with human tumors using the Courtenay clonogenic assay. Stem Cells *1:* 322 (1981).

9 Hoshino, T.; Barker, M.; Wilson, C.B.: The kinetics of cultured human glioma cells. Autoradiographic studies. Acta neuropath. *32:* 235–244 (1975).

10 Hoshino, T.; Wilson, C.B.: Cell kinetic analyses of human malignant brain tumors (gliomas). Cancer *44:* 956–962 (1979).

11 Kohler, G.; Milstein, C.: Continuous cultures of fused cells secreting antibody of predefined specificity. Nature, (Lond.) *256:* 495–497 (1975).

12 Kumar, A.R.V.; Hoshino, T.; Wheeler, K.T.; Barker, M.; Wilson, C.B.: Comparative rates of dead cell removal from brain, muscle, subcutaneous tissue and peritoneal cavity. J. natn. Cancer Inst. *52:* 1751–1755 (1974).

13 Levin, V.A.: Clinical pharmacology of the nitrosoureas; in Prestayko, Nitrosoureas: current status and new developments, pp. 171–180 (Academic Press, New York 1981).

14 Levin, V.A.; Crafts, D.C.; Norman, D.M.; Hoffer, P.B.; Spire, J.P.; Wilson, C.B.: Criteria for evaluating patients undergoing chemotherapy for malignant brain tumors. J. Neurosurg. *47:* 329–335 (1977).

15 Levin, V.A.; Patlak, C.S.; Landahl, H.D.: Heuristic modeling of drug delivery to malignant brain tumors. J. Pharmacokinet. Biopharm. *8:* 257–296 (1980).

16 Levin, V.A.; Wheeler, K.T.: Chemotherapeutic approaches to brain tumors: clinical and experimental observations with dianhydrogalactitol and dibromodulcitol. Cancer Chemother. Pharmacol. *8:* 125–131 (1982).

17 Levin, V.A.; Wright, D.C.; Landahl, H.D.; Patlak, C.S.; Csejtey, J.: In situ drug delivery. Br. J. Cancer *41:* 74–76 (1980).

18 McGuire, W.L.; Carbonne, P.P.; Sears, M.E.; Escher, G.C.: Estrogen receptors in human breast cancer: an overview; in McGuire, Carbonne, Vollmer, Estrogen receptors in human breast cancer, pp. 1–7 (Raven Press, New York 1975).

19 Ozols, R.F.; Willson, J.K.V.; Young, R.C.: Human ovarian cancer colony formation: growth from malignant washings and pharmacologic applications; in Salmon, Cloning of human tumor stem cells, pp. 247–258 (Alan Liss, New York 1980).
20 Rosenblum, M.L.: Chemosensitivity testing for human brain tumors; in Salmon, Cloning of human tumor stem cells, pp. 259–276 (Alan Liss, New York 1980).
21 Rosenblum, M.L.; Deen, D.F.; Levin, V.A.; Dougherty, D.V.; Williams, M.E.; Weizsaecker, M.; Gerosa, M.; Wilson, C.B.: Resistance of brain tumors to chemotherapy. Preliminary studies of rat and human tumors; in Brock, Modern neurosurgery, pp. 104–109 (Springer, Berlin 1982).
22 Rosenblum, M.L.; Dougherty, D.A.: Possible reasons for the failure of brain tumor therapy. Neurosurgery *7:* 296 (1980).
23 Rosenblum, M.L.; Gerosa, M.A.; Dougherty, D.V.: The influence of tumor heterogeneity on chemosensitivity testing of human brain tumors. Stem Cells *1:* 279 (1982).
24 Rosenblum, M.L.; Gerosa, M.A.; Dougherty, D.V.; Reese, C.; Barger, G.R.; Davis, R.L.; Levin, V.A.; Wilson, C.B.: Age-related chemosensitivity of stem cells from human malignant brain tumors. Lancet *i:* 885–887 (1982).
25 Rosenblum, M.L.; Gerosa, M.A.; Wilson, C.B.; Barger, G.R.; Pertuiset, B.; Tribolet, M. de; Dougherty, D.V.: Stem cell studies of human malignant brain tumors. I. Development of the assay and its potential. J. Neurosurg. *58:* 170–176 (1983).
26 Rosenblum, M.L.; Vasquez, D.A.; Hoshino, T.; Wilson, C.B.: Development of a clonogenic cell assay for human brain tumors. Cancer *41:* 2305–2314 (1978).
27 Salmon, S.E.; Alberts, D.S.; Meyskens, F.L.; Durie, B.G.M.; Jones, S.E.; Soehnlen, B.J.; Young, L.; Chen, H.S.G.; Moon, T.E.: Clinical correlations of in vitro drug sensitivity; in Salmon, Cloning of human tumor stem cells, pp. 223–246 (Alan Liss, New York 1980).
28 Salmon, S.E.; Hamburger, A.W.; Soehnlen, B.J.; Durie, B.G.M.; Alberts, D.S.; Moon, T.E.: Quantitation of differential sensitivity of human tumor stem cells to anticancer drugs. New Engl. J. Med. *298:* 1321–1327 (1978).
29a Schnegg, J.F.; Diserens, A.C.; Carrel, S.; Accolla, R.S.; Tribolet, N. de: Human glioma-associated antigens detected by monoclonal antibodies. Cancer Res. *41:* 1209–1213 (1981).
29b Shapiro, J.R.; Shapiro, W.R.: Clonal tumor cell heterogeneity. Prog. exp. Tumor Res., vol. 27, pp. 49–66 (Karger, Basel 1984).
30 Skipper, H.E.; Schabel, F.M., Jr.; Wilcox, W.S.: Experimental evaluation of potential anticancer agents. XXI Scheduling of arabinosylcytosine to take advantage of its S-phase specificity against leukemia cells. Cancer Chemother. Rep. *51:* 125–141 (1967).
31 Stewart, D.J.; Leavens, M.; Maor, M.; Feun, L.; Luna, M.; Bonura, J.; Caprioli, R.; Loo, T.L.; Benjamin, R.S.: Human central nervous system distribution of cis-diaminedichloroplatinum and use as a radiosensitizer in malignant brain tumors. Cancer Res. *42:* 2474–2479 (1982).
32a Tribolet, N. de; Carrel, S.; Mach, J.-P.: Brain tumor associated antigens. Prog. exp. Tumor Res. vol. 27, pp. 118–131 (Karger, Basel 1984).
32b Tveit, K.M.; Endresen, L.; Rugstad, H.E.; Fodstad, Ø.; Pihl, A.: Comparison of two soft-agar methods for assaying chemosensitivity of human tumors in vitro. Malignant melanomas. Br. J. Cancer *44:* 539–544 (1981).
33 Von Hoff, D.D.; Harris, G.J.; Johnson, G.; Glaubiger, D.: Initial experience with

the human tumor stem cell assay system: potential and problems; in Salmon, Cloning of human tumor stem cells, pp. 113–124 (Alan Liss, New York 1980).
34 Weinkam, R.J.; Deen, D.F.: Quantitative dose-response relations for the cytotoxic activity of chloroethyl-nitrosoureas in cell culture. Cancer Res. *42:* 1008–1014 (1982).
35 Weizsaecker, M.; Rosenblum, M.L.: Chemosensitivity testing of human brain tumors in vitro; in Brock, Modern neurosurgery, pp. 144–151 (Springer, Berlin 1982).
36 Wells, J.; Berry, R.J.; Laing, A.H.: The chemosensitivity of freshly explanted tumor cells of various origins as determined by clonal assay and six-day growth in vitro and variation in chemosensitivity with subsequent subculturing; in Dendy, Human tumors in short term culture. Techniques and clinical applications, pp. 158–164 (Academic Press, London 1976).
37 Wheeler, K.T.; Levin, V.A.; Deen, D.F.: The concept of drug dose for in vitro studies with chemotherapeutic agents. Radiat. Res. *76:* 441–458 (1978).
38 Wheeler, K.T.; Tel, N.; Williams, M.; Sheppard, S.; Levin, V.A.; Kabra, P.: Factors influencing the survival of rat brain tumor cells following in vitro treatment with 1,3-bis-(2-chloroethyl)-1-nitrosourea. Cancer Res. *35:* 1464–1469 (1975).

M.L. Rosenblum, MD, Brain Tumor Research Center of the Department of Neurological Surgery, School of Medicine, University of California, San Francisco, CA 94143 (USA)

Prog. exp. Tumor Res., vol. 28, pp. 18–31 (Karger, Basel 1984)

Human Brain Tumor Xenografts

S. C. Schold, Jr.[a], *H. S. Friedman*[b]

[a]Departments of Medicine (Neurology), Pathology (Neuropathology), and
[b]Pediatrics, Duke University Medical Center, Durham, N.C., USA

Introduction

Heterologous transplantation of human tumor tissue into animals offers the experimental oncologist the chance to study human neoplastic tissue directly under controlled conditions. The defense mechanisms of immunologically intact animals limit the success of such procedures, and even though the relationships between the systemic immune system and the brain are complex, reliable heterotransplantation of neoplasms in the central nervous system appears to require some degree of immunodeficiency in the recipient animals. The development of the congenitally athymic 'nude' mouse and refinement of techniques for the production of acquired immunodeficiencies have permitted more detailed study of a variety of heterotransplanted human tumors, including brain tumors. In this chapter, we review the experience with xenografts of human brain neoplasms, including studies of their morphology, growth characteristics, and therapeutic sensitivity.

History of Human Brain Tumor Heterotransplantation

Transplantation into Immunocompetent Hosts

Greene and a number of coworkers, in the 1940s and 1950s, demonstrated that human glioblastomas and some other primary central nervous system neoplasms could be successfully transplanted into immunologically privileged sites of normal heterologous hosts. *Greene and Arnold* [13] reported in 1945 that 3 of 3 human glioblastomas grew in the anterior chamber of the guinea pig eye after transplantation. They were surprised by the long latency period before growth (often more than 200 days); but once growing, the tumors resembled human glioblastomas histologically, and at least 1 was serially transplantable. *Krementz and Greene* [18] extended this series in 1953 when they described 9 of 22 successful glioblastoma transplants into the anterior chamber of the guinea pig eye. In this series, up to 14 months were required for successful growth, and, again, histologic similarity to the parent tumors was maintained. They also reported growth of 2 of 3 human medulloblastomas. *Greene* later performed some novel experiments using this system. He attempted to isolate and serially transplant the neovascular component of these tumors. He claimed to have produced 'sarcomas' by this method and suggested that he had transplanted truly neoplastic endothelial cells. Because reliable cell markers were not available to him at the time, these data are difficult to interpret, but the approach was an interesting one, and the question about the nature of the endothelial cell in the glioblastoma remains compelling.

Manuelidis [19] also reported successful transplantation of human glioblastomas into the guinea pig eye anterior chamber. In a morphologic study, he described 6 of 19 successful transplants, including 3 serially transplantable tumor lines. During early passage, the tumors resembled the multiform appearance of the parent tumor, but as serial transplantation progressed, increasing morphologic uniformity appeared. *Manuelidis* [19] suggested that this change represented selection of a subpopulation of the original tumor, rather than the effect of the novel environment.

Another immunologically privileged site, the brain itself, was used for xenograft experiments by *Greene* [14]. He reported that 8 of 11 human glioblastomas transplanted into guinea pig brains showed progressive growth. The average latency to the appearance of symptoms in the recipient animals was 4 months, and once again, the tumors in the animals resembled human glioblastomas and were successfully transplanted 'whenever attempted'.

Martin [20], however, in 1951, reported a lack of success in transplanting human glioblastomas into the anterior chamber of guinea pig eye. He did describe 1 successful transplant, but unfortunately the total number of tumors transplanted and the periods of observation were not reported. In another study, preconditioning of rats with irradiation and corticosteroids did not improve the success in growing tumors from intracerebrally transplanted human glioblastomas [22]. *Wilson and Barker* [34] attempted intrathecal transplantation of primary human brain tumors into guinea pigs. Only 1 of 20 tumors, a medulloblastoma, grew, whereas all 6 glioblastomas failed to grow. They concluded that the cerebrospinal fluid was not advantageous as a growth medium for human brain tumor xenografts.

Greene's success in the transplantation of human neoplasms into immunocompetent heterologous hosts has never been achieved in another laboratory, and the underlying biologic factors affecting successful transplantation have never been fully explored.

Transplantation into Immunoincompetent Hosts

The discovery of the mutant athymic 'nude' mouse in the 1960s [24] simplified the problem of heterologous transplantation of human tumors by affording a high rate of successful transplantation without prior conditioning of the hosts [9, 28]. Hundreds of human tumors have been transplanted into these animals with significant but variable 'take rates', depending primarily on the histologic tumor type, but also on the endocrine status of the animals, their health, and the transplantation techniques. *Rana* et al. [25] reported the first successful transplantation of a human glioblastoma into nude mice. The tumor grew in 3 of 6 mice with 2- to 3-month latencies to growth, and it resembled morphologically the original surgical specimen. *Wara* et al. [33] described 2 successful glioblastoma heterotransplantations, and *Reid* et al. [27] described another. Subsequent work from Sloan-Kettering Cancer Center and Duke University expanded the series of human glioblastomas transplanted into nude mice. Over 40 such tumors have now been reported. Both laboratories have found a very high rate of successful transplantation. *Shapiro* et al. [30] initially reported 7 of 7 subcutaneous and 6 of 7 intracerebral takes. In a later series they described 16 of 21 successful intracerebral transplantations of 'human malignant glial tumors', including 14 of 17 glioblastomas and 2 of 2 'malignant astrocytomas'. One gliosarcoma and 1 'grade 2' astrocytoma failed to grow [15]. The growth rate of the initial transplantations could not be determined because of the intracerebral location of the tumors, but recipient animals died between 62 and 239 days after the transplantation. *Schold* et al. [29] reported successful subcutaneous transplantation of 16 of 17 human glioblastomas with highly variable latencies to appearance and growth rates. Some individual tumors showed evidence of growth almost immediately after transplantation, whereas others required over 6 months and, in one instance, over 300 days for initial growth to become evident. Volume-doubling times of these subcutaneous masses varied from 4 to 30 days. This successful transplantation rate of

Table I. Biologic characteristics of the human medulloblastoma tumor line TE-671

In vitro	
Population doubling time, h	31.5
Colony-forming efficiency, % (agarose medium)	9.0 ± 3.0
Saturation density, cells/cm^2	3.98×10^5
In vivo (athymic mice) subcutaneous tumors	
Doubling time, days	3.2 ± 1.0
Latency to 500 mm^3, days	20.7 ± 1.5
Cell cycle time (T_C), h	24
S phase (T_s), h	14
G_1 phase (T_{G1}), h	7
G_2 phase (T_{G2}), h	3

over 80% exceeds that for any other reported human tumor type in nude mice, presumably reflecting an important biologic characteristic of these tumors. They can be transplanted either intracerebrally or subcutaneously, but the animals must be observed for many months before conclusions can be reached about the success of the transplant.

A series of established cell lines derived from human gliomas also has been transplanted into nude mice. Surprisingly, only 4 of 15 permanent glioma lines formed transplantable subcutaneous tumors [3], but 3 of these lines have been serially transplanted and extensively characterized [8]. The lack of tumorigenicity of some permanent human tumor cell lines emphasizes the ambiguity of the concept of a neoplastic cell, especially as it appears that human gliomas are more readily established as tumor lines in nude mice than they are in culture.

One other important brain tumor line has now been characterized extensively in athymic mice. Cell line TE-671 was originally established in tissue culture by *McAlister* et al. [21] from a human medulloblastoma. It is the only established human medulloblastoma cell line available, and both its in vitro biologic characteristics and its behavior in nude mice have now been determined (table I) [10]. It grows subcutaneously in nude mice, with a volume-doubling time of 3.2 days, and the median survival of animals receiving 5 μl of a tumor suspension intracerebrally is 25.0 days.

Curiously, none of a large number of other histologic brain tumor types has consistently grown in athymic mice. Anaplastic astrocytomas, ependymomas, astrocytomas, and oligodendrogliomas have not grown

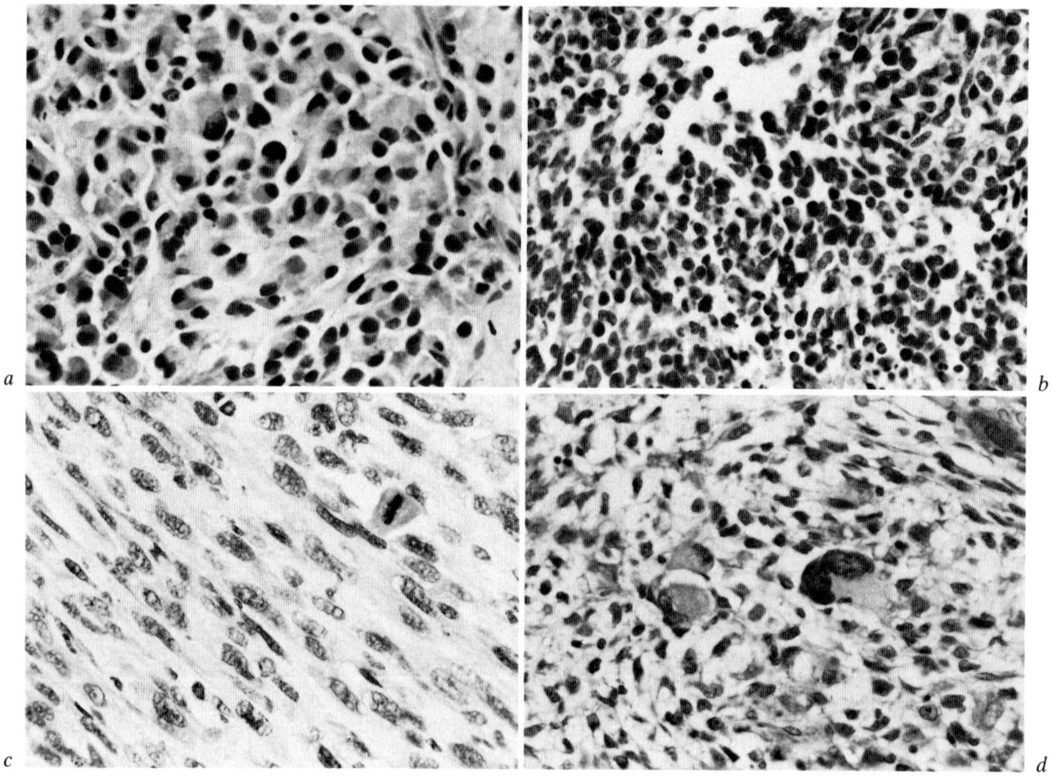

Fig. 1. A variety of cell types is present in the human glioblastomas growing subcutaneously in athymic mice. *a* Astrocytes. *b* Small anaplastic cells. *c* Spindle cells. *d* Giant cells. HE. ×245.

except in a few instances, and even in those cases, they were not serially transplantable. In a sense, growth in nude mice is, therefore, a reflection of the clinical behavior of the tumor, although apparently no more so than is its histologic appearance.

Characterization of Human Brain Tumor Xenografts

Morphologic and Immunohistochemical Studies

Morphologic and immunohistochemical descriptions of these transplanted glioblastomas have been of some interest. The tumors grow as

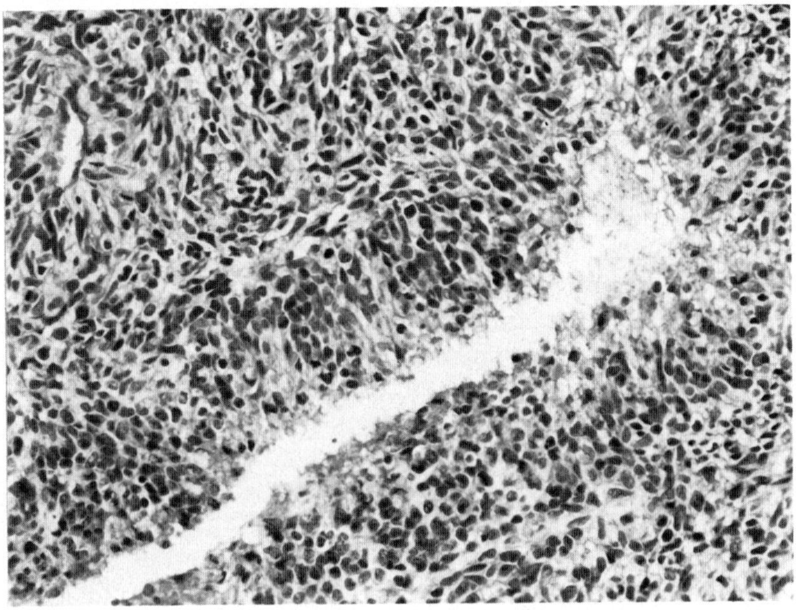

Fig. 2. Necrosis with pseudopalisading is frequently present in the heterotransplanted human glioblastomas. HE. ×240.

lobulated, confined masses in the subcutaneous space and as either infiltrative or discrete masses in the brain. In general, the transplants have appeared to maintain cytologic resemblance to neoplastic glia; but in many instances the transplanted tumors have been more uniform than the parent tumors [15, 29]. Whether this represents selection, sampling, or the effect of the novel environment remains to be determined, although presumably some degree of cellular selection occurs in the transplantation procedure.

The full range of cytologic features of human glioblastomas has been observed in the transplanted tumors, including astrocytes, small anaplastic cells, spindle cells, and giant cells (fig. 1). Necrosis with pseudopalisading is common, but, curiously, endothelial proliferation is not (fig. 2).

In some instances, serial transplantation of human glioblastomas in nude mice has appeared to produce morphologic changes. *Wara* et al. [33] first described this phenomenon in a serially transplanted glioblastoma that became progressively more 'sarcomatous' morphologically. In the Duke University series, the initial astrocytic cell population of one tumor line (N-456) was gradually replaced in serial transplantation by a homogeneous population of anaplastic spindle cells. However, in a number of other tumor lines in this series, the original morphology was maintained in serial passage. In the series from Duke University, 12 of 14 transplanted glioblastomas expressed glial fibrillary acidic protein in nude mice, the amount expressed varying from under 5 to over 75% of cells in individual tumor lines. Glial fibrillary acidic protein expression was usually maintained in serial animal passage, although in two instances it was lost by the 5th passage. Fibronectin was expressed by 7 of 8 transplanted glioblastomas in nude mice and was detectable in the blood of mice bearing U-118 MG, a glioma line that expressed fibronectin in vitro [16].

The morphology of the tumors derived from permanent cell lines has also varied, although each has remained recognizable and distinct in serial animal passage. The medulloblastoma cell line contains small, dark, anaplastic cells with pseudorosette formation and is apparently similar to the tumor from which it was derived.

Karyology. Karyotypic analysis of a number of nude mouse-borne human glioblastomas has now been accomplished [4]. In both this series and that from *Bradley* et al. [7] on human gliomas in immunosuppressed animals a human karyotype was retained in the transplanted tumors. Aneuploidy was also preserved in the transplants if it was present in the parent tumor. Furthermore, as described by *Bigner and Mark* [4], marker chromosomes that were present in the parent tumor were also often present in the transplants.

The karyotype of the medulloblastoma cell line in vitro and in vivo reveals modal chromosome counts in the near-tetraploid range. The karyotyped cells in all preparations are very similar, but as there are numerical variations from cell to cell, there is no true stem karyotype. These near-tetraploid cells closely resemble a doubled complement of the chromosome set illustrated by *McAllister* et al. [21] in their original report of this cell line.

Antigenicity. Bourdon et al. [6] have reported that 6 of 8 human glioblastoma xenografts growing subcutaneously in nude mice express the glioma-mesenchymal extracellular matrix antigen 81C6. This antigenic marker, which is generally confined to the areas of proliferative endothelium in anaplastic glioma biopsies, is present in the transplanted mouse-borne tumors throughout the tumor stroma. Its significance is as yet unknown. The expression in nude mouse-borne glioblastomas of additional antigens defined by brain and glioma-derived monoclonal antibodies is now being evaluated.

Therapeutic Implications of Human Brain Tumor Xenografts

There exist now a number of reports on the effect of chemotherapy and radiation therapy on human gliomas growing in nude mice. Both directly transplanted tumors and those derived from established human glioma lines have been used; furthermore, tumors have been evaluated when growing in both subcutaneous and intracerebral locations. Each of these systems has its strengths and can be justified, depending on the nature of the experiment. Direct transplants of biopsy specimens have more relevance for individual patients, although these studies are limited by the long period of tumor growth required before therapeutic testing can be done. In contrast, tumors derived from established cell lines have the advantages of being more uniform and of being well characterized in many respects. These latter tumors clearly have less relevance to the patient from whom they were derived than have the direct transplants. Nevertheless, tumors developed from established cell cultures should be more relevant to the study of human brain neoplasms than are many commonly used animal tumor models, as they were originally obtained from human brain tumors and are neoplastic by definition.

The intracerebral tumors more closely mimic the clinical situation; the presence of the blood-brain barrier makes drug delivery an important concern, and a single well-defined end-point, survival, is usually evaluated. Alternatively, the subcutaneous tumors are accessible for continuous measurements, and their response can, therefore, be analyzed in a number of ways. More important is that the extracerebral location allows the investigator to focus on the question of the cellular sensitivity of these neoplasms with less concern about the drug delivery problem. Obviously, if a tumor cell is insensitive to a particular agent, improving delivery of that agent into the brain will not accomplish a great deal. If a tumor has demonstrated cellular sensitivity to an agent in the subcutaneous space, however, the drug or the technique of its delivery can be modified to bring about intracerebral activity.

Wara et al. [33] first demonstrated semiquantitative effects of 1,3-bis(2-chloroethyl)-1-nitrosourea (BCNU) and radiation on a subcutane-

Table II. Treatment of the human glioma cell line D-54 MG growing subcutaneously in athymic mice

Drug	Dose schedule mg/m^2	Growth delay	Number of tumor regressions/number of tumors treated
BCNU	75×1	14.3*	16/20*
Procarbazine	700 daily × 5	16.0*	19/20*
AZQ	6.75 daily × 5	14.9*	14/20*
Cyclophosphamide	516 daily × 2	22.3*	10/10*
Cis-platinum	32.5 × 1	8.4*	12/20*
Mithramycin	1.3 daily × 5	1.7	2/30
Methotrexate	24 daily × 5	0.1	1/10
BW-301U	70 daily × 4	0.8	0/10
BCNU + procarbazine	37.5 × 1 + 700 daily × 5	>100*	21/21*

*$p<0.05$.

ous human glioma growing in nude mice. *Shapiro* et al. [30] reported in detail on the response of 2 human glioma xenografts growing intracerebrally in nude mice to nitrosoureas and procarbazine and briefly on the responses of 3 additional tumors [31]. They demonstrated the sensitivity of 1 tumor (S.R.) to BCNU and its resistance to procarbazine, whereas a second transplant (L.M.) was unaffected be either BCNU or 1-(2-chloroethyl)-3-(2,6-dioxo-3-piperidyl)-1-nitrosourea (PCNU), but was quite sensitive to procarbazine. *Bullard* et al. [8] treated 3 human glioma-derived cell lines in nude mice with BCNU. They also found variable sensitivity to this agent. *Schold and Bigner* [29] extended these studies by documenting variable responses of 5 human glioma tumor lines (3 cell line derived and 2 direct transplants) to 3 agents that have been used to treat human brain tumors: BCNU, procarbazine, and mithramycin. Reflecting the clinical experience, mithramycin was ineffective, whereas BCNU and procarbazine were each active against at least some of the glioma tumor lines. One of these lines, D-54 MG, has been studied in greater detail (table II). It is highly sensitive to a BCNU-procarbazine combination, whereas cis-platinum, aziridinylbenzoquinone (NSC-183986; AZQ), and cyclophosphamide, as well as BCNU and procarbazine individually, have only moderate activity. Mithramycin, methotrexate, and a lipid-soluble antifolate agent, BW-301 U, were inactive. Intracerebral D-54 MG is also

Table III. Treatment of intracerebral D-54 MG in athymic mice with four chemotherapeutic agents

Drug	T/C[1]
BCNU	243, 196
Procarbazine	>700
AZQ	171, 172
Cis-platinum	160, 115

[1]Median survival of treated animals/median survival of control animals × 100. Each value was statistically significant at $p<0.01$ by the Wilcoxon rank-sum test.

Table IV. Treatment of the human medulloblastoma cell line TE-671 growing subcutaneously in athymic mice

Drug	Dose schedule mg/m^2	Growth delay	Number of tumor regressions/number of tumors treated
Cyclophosphamide	516 daily × 2	16.5*	9/9*
Vincristine	7.3 × 1	15.0*	3/8
AZQ	6.75 daily × 5	3.3*	1/10
Cis-platinum	25 × 1	3.0*	1/10
Methotrexate	24 daily × 5	2.8	2/10
BCNU	75 × 1	1.4	1/10
Procarbazine	700 daily × 5	0.8	2/8
BW-301U	70 daily × 4	0.2	1/8
Radiation	2,500 rad (single fraction)	39.5*	7/9*
	1,500 rad (single fraction)	21.1*	2/10

*$p<0.05$.

sensitive to procarbazine, BCNU, AZQ, and cis-platinum, in that order (table III) [26]. *Slagel* et al. [32] have also determined the sensitivity of U-251 MG (BN-3) to radiation, procarbazine, and AZQ. They found that an irradiation-chemotherapy combination was more effective than was either treatment alone.

The sensitivity of the medulloblastoma line in mice to a number of chemotherapeutic agents is now known as well. It is resistant to standard brain tumor agents, such as BCNU and procarbazine; it shows slight sensitivity to AZQ, and it is quite sensitive to cyclophosphamide, vincristine, and radiation (table IV) [11]. *Slagel* et al. [32] have also reported resistance of TE-671 (BN-2) in nude mice to several agents, but this resistance was significantly affected by an irradiation-chemotherapy combination. As there is no generally accepted experimental model of medulloblastoma that is suitable for therapeutic studies, this tumor line in mice may be extremely valuable for preclinical studies of potentially active agents.

Most of the drugs tested against human brain tumors in nude mice have an established clinical track record, as is documented where in this volume. The nitrosoureas and procarbazine have limited but definite activity against human brain tumors, whereas mithramycin is ineffective. The antifolates are inactive. The status of cis-platinum is uncertain, although it appears to have some activity. In general, therefore, therapeutic results in xenografts reflect clinical experience and justify the use of this model to evaluate potential therapeutic approaches.

AZQ, for example, is a drug of considerable current interest in brain tumor chemotherapy because of its activity in a number of animal brain tumor models [17]. We have tested AZQ against a series of 7 human brain tumor lines in nude mice, and we have once again found variable sensitivity to the drug. Statistically significant growth delays were seen with 4 lines (D-54 MG, TE-671, N-735, and N-519), whereas U-118 MG, U-251 MG, and N-456 were insensitive. On this basis, it appears that the drug will find a place in the treatment of human gliomas, although the basis for sensitivity and resistance toit is not yet known. Cyclophosphamide also has activity against both subcutaneous and intracerebral human brain tumors in nude mice. Growth delays of 22.3 days in D-54 MG and 16.5 days in TE-671 are highly significant and compare favorably with, for example, BCNU and procarbazine. Cyclophosphamide has not undergone extensive clinical trial against primary human central nervous system tumors, because the drug does not penetrate the intact blood-brain barrier and because it was minimally active in standard screening systems [12]. Our results indicate, however, that it has activity against human glioma and medulloblastoma lines and that it should be reevaluated clinically. If the barrier is a practical clinical problem, it might be circumvented by high-dose intravenous treatment [1].

Large clinical trials of chemotherapy of human brain tumors are difficult, time-consuming, and expensive. The choice of agents to be tested in phase II trials is relatively haphazard and often based on limited experimental data. An efficient screening system using human brain tumors is theoretically attractive and should increase the yield of the clinical studies.

Future Prospects

Human gliomas grow reliably in athymic mice, demonstrate biologic traits that correspond to the parent tumors, and have demonstrated variable sensitivity to a number of standard chemotherapeutic agents. These systems have been developed to allow testing of newer agents of potential interest, such as AZQ, BW-301U, and dianhydrogalactitol. We are now at the point where agents in the preclinical stages of development can be tested as a secondary screen for potentially active drugs [2]. Further refinement of the system, such as by kidney capsule implantations [5] or preconditioning of the animals [23], might allow more rapid characterization of individual tumors. However, nude mouse screening systems are expensive and much more time-consuming than are the systems that use the standard animal tumor models, such as the intracerebral L1210 leukemia and the murine ependymoblastoma.

Although on theoretical grounds one could anticipate that the nude mouse system will more accurately reflect the clinical situation, developments in the next few years will demonstrate whether the investment is justified.

Acknowledgments

The work described in this chapter was done at Duke University and was supported by Grant No. NS-16974 from the National Institute of Neurological and Communicative Diseases and Stroke, National Institutes of Health Grant No. 5T32AG 00007, and American Cancer Society Grant No. CH-191. Dr. *Sandra H. Bigner* prepared the photomicrographs. Drs. *Joseph Pasternak* and *Dennis Groothuis* did the kinetic studies on the medulloblastoma cell line in athymic mice. The Burroughs-Wellcome Co. supplied compound BW-301U.

References

1 Allen, J.C.; Helson, L.: High-dose cyclophosphamide chemotherapy for recurrent CNS tumors in children. J. Neurosurg. *55:* 749–756 (1981).

2 Bellett, R.E.; Danna, V.; Mistrangelo, M.S.; Berd, D.: Evaluation of a 'nude' mouse-human tumor panel as a predictive secondary screen for cancer chemotherapeutic agents. J. natn. Cancer Inst. *63:* 1185–1188 (1979).

3 Bigner, D.D.; Markesberry, W.R.; Pegram, C.N.; Westermark, B.; Ponten, J.: Progressive neoplastic growth in nude mice of cultured cell lines derived from human gliomas (Abstract). J. Neuropath. exp. Neurol. *36:* 593 (1977).

4 Bigner, S.H.; Mark, J.: Chromosomes and choromosomal progression of human gliomas in vivo, in vitro and in athymic nude mice; in Rosenblum, Wilson, Prog. exp. Tumor Res., vol. 27, pp. 67–82 (Karger, Basel 1984).

5 Bogden, A.E.; Cobb, W.R.; Lepage, D.J.; Haskell, P.M.; Gulkin, T.A.; Ward, A.; Kelton, D.E.; Esber, H.J.: Chemotherapy responsiveness of human tumors as first transplant generation xenografts in the normal mouse: six-day subrenal capsule assay. Cancer *48:* 10–20 (1981).

6 Bourdon, M.A.; Wikstrand, C.J.; Furthmayr, H.; Bigner, D.D.: A human glioma-mesenchymal extracellular matrix antigen defined by monoclonal antibody. Cancer Res. *43:* 2796–2805 (1983).

7 Bradley, N.J.; Bloom, H.J.G.; Davis, A.J.S.; Swift, S.M.: Growth of human gliomas in immune-deficient mice: a possible model for preclinical therapy studies. Br. J. Cancer *38:* 263–272 (1978).

8 Bullard, D.E.; Schold, S.C.; Bigner, S.H.; Bigner, D.D.: Growth and chemotherapeutic response in athymic mice of tumors arising from human glioma-derived cell lines in athymic mice. J. Neuropath. exp. Neurol. *40:* 310–327 (1981).

9 Fogh, J.; Tiso, J.; Orfeo, T.; Sharkey, F.E.; Daniels, W.P.; Fogh, J.M.: Thirty-four lines of six human tumor categories established in nude mice. J. natn. Cancer Inst. *64:* 745–751 (1980).

10 Friedman, H.S.; Bigner, D.D.; McComb, R.D.; Schold, S.C., Jr.; Bigner, S.H.: A model for human medulloblastoma growth, morphology, and chromosomal analysis in vitro and in athymic mice J. Neuropath. exp. Neurol. (in press).

11 Friedman, H.S.; Schold, S.C., Jr.; Varia, M.; Bigner, D.D.: Chemotherapy and radiation therapy of human medulloblastoma in athymic nude mice. Cancer Res. (in press).

12 Geran, R.I.; Congleton, G.F.; Dudeck, L.E.; Abbott, B.J.; Fargus, J.L.: A mouse ependymoblastoma as an experimental model for screening potential antineoplastic drugs. Cancer Chemother. Rep. *4:* 53–87 (1974).

13 Greene, H.S.N.; Arnold, H.: The homologous and heterologous transplantation of brain and brain tumors. J. Neurosurg. *2:* 315–331 (1945).

14 Greene, H.S.N.: The transplantation of human brain tumors to the brains of laboratory animals. Cancer Res. *13:* 422–426 (1953).

15 Horten, B.C.; Basler, G.A.; Shapiro, W.R.: Xenograft of human malignant glial tumors into brains of nude mice: a histopathological study. J. Neuropath. exp. Neurol. *40:* 493–511 (1981).

16 Jones, T.R.; Ruoslahti, E.; Schold, S.C., Jr.; Bigner, D.D.: Fibronectin and glial fibrillary protein expression in normal human brain and anaplastic human gliomas. Cancer Res. *42:* 168–177 (1982).

17 Khan, A.H.; Driscoll, J.S.: Potential central nervous system antitumor agents. Aziridinylbenzoquinones. J. med. Chem. *19:* 313–317 (1976).

18 Krementz, E.T.; Greene, H.S.N.: Heterologous transplantation of human neural tumors. Cancer *6:* 100–110 (1953).

19 Manuelidis, E.E.: The fate of serial heterologously transplanted glioblastoma multiforme in the eye of the guinea pig. Am. J. Path. *48:* 65–89 (1966).

20 Martin, J.: The transplantation of human brain tumors into animal hosts. J. Neuropath. exp. Neurol. *10:* 40–47 (1951).

21 McAllister, R.M.; Isaacs, H.; Rongey, R.; Peere, M.; Au, W.; Soulcup, S.W.; Gardner, M.B.: Establishment of a human medulloblastoma cell line. Int. J. Cancer *20:* 206–212 (1977).

22 Miyawaki, H.; Ishii, S.: The heterologous intracerebral transplantation of human brain tumors. Archs Path. *70:* 508–519 (1960).

23 Ohsugi, Y.; Gershwin, M.E.; Owens, R.B.; Nelson-Rees, W.A.: Tumorigenicity of human malignant lymphoblasts: comparative study with unmanipulated nude mice, antilymphocyte serum-treated nude mice, and x-irradiated nude mice. J. natn. Cancer Inst. *65:* 715–718 (1980).

24 Pantelouris, E.M.: Absence of thymus in a mouse mutant. Nature *217:* 370–371 (1968).

25 Rana, M.W.; Pinkerton, H.; Thornton, H.; Nagy, D.: Heterotransplantation of human glioblastoma multiforme and meningioma to nude mice. Proc. Soc. exp. Biol. Med. *155:* 85–88 (1977).

26 Rawlings, C.E., III; Schold, S.C., Jr.; Bigner, S.H.; Bigner, D.D.: Growth and treatment of an intracerebral human glioma tumor line in athymic mice with BCNU, procarbazine, AZQ, and cis-platinum. Neurosurgery (in press).

27 Reid, L.M.; Holland, J.; Jones, C.; Wolf, B.; Niwayama, G.; Williams, R.; Kaplan, N.O.; Sato, G.: Some of the variables affecting the success of transplantation of human tumors into the athymic nude mouse; in Houchens, Ovejera, The use of athymic (nude) mice in cancer research, pp. 107–121 (Fischer, New York 1978).

28 Rygaard, J.; Povlsen, C.O.: Heterotransplantation of a human malignant tumor to 'nude' mice. Acta path. microbiol. scand. *77:* 758–760 (1969).

29 Schold, S.C., Jr.; Bigner, D.D.: Treatment of five subcutaneous human glioma tumor lines in athymic mice with BCNU, procarbazine, and mithramycin. Cancer Treat. Rep. (in press).

30 Shapiro, W.R.; Basler, G.A.; Chernik, N.L.; Posner, J.B.: Human brain tumor transplantation into nude mice. J. natn. Cancer Inst. *62:* 447–453 (1979).

31 Shapiro, W.R.; Basler, G.A.: Chemotherapy of human brain tumors transplanted into nude mice; in Paoletti, Walker, Butti, Knerich, Multidisciplinary aspects of brain tumor therapy, pp. 309–316 (Elsevier, Amsterdam 1979).

32 Slagel, D.G.; Feola, J.; Houchens, D.P.; Ovejera, A.A.: Combined modality treatment using radiation and/or chemotherapy in an athymic (nude) mouse-human medulloblastoma and glioblastoma xenograft model. Cancer Res. *42:* 812–816 (1982).

33 Wara, W.M.; Begg, A.; Phillips, T.L.; Rosenblum, M.L.; Vasquez, D.; Wilson, C.B.: Growth and treatment of human brain tumors in nude mice – preliminary communication; in Houchens, Ovejera, The use of athymic (nude) mice in cancer research, pp. 251–256 (Fischer, New York 1978).

34 Wilson, C.B.; Barker, M.: Intrathecal transplantation of human neural tumors to the guinea pig. J. natn. Cancer Inst. *41:* 1229–1240 (1968).

S.C. Schold, Jr., MD, Department of Medicine (Neurology), Duke University Medical Center, Durham, NC 27706 (USA)

Prog. exp. Tumor Res., vol. 28, pp. 32–50 (Karger, Basel 1984)

Novel Modes of Chemotherapy Administration

David J. Stewart

Ontario Cancer Foundation, Ottawa Clinic, General Hospital Division, Ottawa, Ont., Canada

Most patients with intracerebral tumors still die less than 2 years after diagnosis [17]. To meet the challenge that these tumors present, new therapeutic modalities are urgently needed. During the 1950s and 1960s, many studies were conducted to evaluate the possible efficacy of intracarotid chemotherapy infusions for treating intracerebral gliomas. Interest in this drug delivery mode waned because of technical difficulties, complications, and disappointing results; but in recent years there has been a revival of interest in this approach. I will discuss the rationale for and utility of the intra-arterial route for chemotherapy administration, and describe briefly the intrathecal and intratumoral routes of drug delivery.

Intra-Arterial Route

Pharmacological Considerations

Under certain conditions, the intra-arterial administration of chemotherapy may have advantages over intravenous and oral routes of drug administration. It affords a marked augmentation of the peak arterial concentration of drug (Co), and a greater period of time over which tumor is exposed to the maximum concentration of drug, as depicted by the area under the concentration-time curve (C×t) [15]. As many chemotherapeutic agents have a steep dose-response curve, this enhanced exposure to the drug may result in greater antitumor efficacy. Once the intra-arterial infusion ends, however, local arterial drug concentrations and systemic drug concentrations rapidly equilibrate [3, 15].

Many factors influence the uptake of drug by tissue during intra-arterial infusion. A decrease in the regional blood flow markedly enhances the arterial Co and C×t, and somewhat decreases systemic exposure to the drug [3, 9, 15]. Balloon catheters [22] and embolization by microspheres [29] have been used to decrease flow in extracerebral arteries. Although the application of these techniques is limited for brain tumors by the attendant risk of cerebral ischemia, balloon catheters have been used with good effect for selective infusion of 1,3-bis(2-chloroethyl)-1-nitrosourea (BCNU) into brain tumors [54]. Transient external compression of the carotid or vertebral arteries has also been used in conjunction with intra-arterial chemotherapy [37].

Rapid intravascular degradation of the drug used for intra-arterial chemotherapy decreases systemic exposure to the drug and thereby decreases systemic toxicity [3, 15]. Because of their rapid breakdown, the nitrosoureas and highly labile alkylating agents have been recommended for intra-arterial use [3]. However, the factor limiting the application of intracarotid infusions is generally local, rather than systemic, toxicity. The extraction fraction (the rate of blood-brain exchange) and the tissue's partition coefficient (a measure of a tissue's capacity to take up the drug) are of less importance than is the rate of drug breakdown for promoting a possible advantage of intra-arterial administration of chemotherapeutic agents over their intravenous administration [15].

Rapid infusion of the drug will result in a higher arterial Co than will slow infusion, but the C×t is not affected by the infusion rate [9, 15]. The C×t is usually considered to be more important than Co in determining drug efficacy [15]. Since a high Co might increase the chances of toxicity [14, 33, 61] it can be suggested that the therapeutic index may be better for slow than for rapid drug infusion techniques. Protein binding, membrane transport, and extracellular diffusion also affect the interaction between drug and tissue, but their influence on drug uptake is probably the same after intra-arterial and intravenous administration [3].

Because infused drug does not immediately mix completely with blood in the infused vessel, distribution of drug to the tumor may be uneven [37]. Reducing the blood flow rate could result in more thorough mixing, as suggested earlier. In some patients, the artery that is infused may not supply the entire tumor. For hepatic artery infusions, it is possible to predict drug distribution patterns using radioimaging techniques [27]; such an approach to identify the region(s) of the tumor that are infused has not been explored for intracerebral lesions.

No benefit would be anticipated from the intra-arterial infusion of drugs that require hepatic activation for antineoplastic activity, such as cyclophosphamide.

Pharmacokinetics of Individual Drugs

We have observed that the pharmacokinetics for cisplatin following carotid artery infusions were similar to those measured following cisplatin infusions administered intravenously and into other arteries, although Co values were higher with carotid infusions because of a more rapid infusion rate [64]. After cisplatin arterial infusions, regional veins had higher platinum Co and C×t values than did peripheral veins. This finding has since been confirmed by others [8]. Platinum concentrations in human tumors also appear to be higher after intra-arterial infusions than after intravenous infusions [62, 67]. In dogs, platinum concentrations in cerebrospinal fluid (CSF) and brain are enhanced by intracarotid infusion [38]. Moreover, we have observed a glioblastoma that had progressed during treatment with intravenously administered cisplatin that responded to intra-arterial cisplatin administration [*Stewart,* unpubl. data, 1983]. Other tumor types that usually are refractory to cisplatin administered intravenously may respond to intra-arterial administration [7, 35].

In an attempt to improve the therapeutic index of intra-arterial cisplatin, sodium thiosulfate has been given intravenously concurrently with the intra-arterial administration of cisplatin. The sodium thiosulfate binds and inactivates cisplatin that might otherwise attain high concentrations in kidney. Following sodium thiosulfate administration, cisplatin concentrations in peripheral veins decrease and clearance increases by 14% [8]; studies with thiosulfate have not been performed in conjunction with intracarotid cisplatin administration.

Experiments with BCNU in monkeys have shown that the concentration of the drug in brain is markedly augmented by intracarotid, as compared to intravenous, infusion [36]. Neocarzinostatin has been used intra-arterially because it has an extremely short half-life, is more toxic to glioblastoma cells than to glial cells, and leaks rapidly from capillaries [41].

Hepatic artery infusions of doxorubicin (Adriamycin®) result in higher hepatic concentrations than do venous infusions [28]. A computer model predicts that doxorubicin infusion into extrahepatic arteries would result in increased local exposure as compared to intravenous infusion, but no decrease in systemic exposure to the drug [9]. Although no doxorubicin was found in animal brain following intracarotid infusion of the drug alone, other studies have shown that intracarotid administration of the drug after disruption of the blood-brain barrier (BBB) by intracarotid mannitol results in high doxorubicin concentrations in brain and severe neurotoxicity [47].

For bleomycin, a computer model predicted only slight improvement in efficacy and no decrease in toxicity with intra-arterial administration [9]. The model also predicted both increased activity and decreased toxicity of 5-fluorouracil (5-FU) and decreased toxicity of melphalan. No pharmacokinetic data are available on the intracarotid infusion of these three agents. In the case of methotrexate (MTX), intra-arterial infusion should be advantageous, according to the computer model; but only slight augmentation of tissue concentrations are actually found [9]. Pretreatment with the intracarotid administration of mannitol augments the penetration of MTX into brain, and possibly may augment its penetration into intracerebral tumors [46], although this remains controversial [1].

Intracarotid injection of etoposide resulted in higher concentrations in canine brain and CSF and lower concentrations in extracerebral organs than did intravenous injections of the drug [58]. Etoposide appeared to be either extensively metabolized or tissue-bound [58]. Similar results were obtained with aziridinylbenzoquinone (AZQ) [25].

Choice of Drugs

The chemical properties of the drug solution should be such that blood vessels and normal brain are not damaged. A drug with a high extraction fraction and partition coefficient and rapid spontaneous breakdown would offer some additional advantage, as mentioned earlier. Previously it was thought that the ability of the agent to cross the intact BBB was also essential for it to be active against intracerebral tumors. It is now clear that this is not so. For example, cisplatin penetrates poorly into normal brain and CSF, but concentrations of the drug that appear comparable to those attainable in extracerebral tumor are found in human intracerebral tumors after intravenous administration [67], and intravenous administration of cisplatin has produced regression of glioblastomas and of several types of brain metastases [30, 32]. In addition, combinations containing lipid-insoluble drugs have produced response rates for brain metastases from breast cancer that are identical to those achieved in cases of extracerebral metastases [57]. How, and to what extent, a drug's ability to penetrate into normal brain determines its activity against intracerebral tumors remains unknown. The BBB may be largely disrupted within intracerebral tumors [77], although the degree of disruption probably varies from one tumor to another [24]. To confirm this, we have demonstrated high human intracerebral tumor concentrations of many drugs including cisplatin [62, 63, 67], 3-deazauridine [65], phosphonacetyl-*L*-aspartate [66], vinblastine [63], m-AMSA [81], methylglyoxyl-bis-(guanylhydrazone) [56], AZQ [59], etoposide [68], and pentamethylmelamine [*Stewart,* unpubl. data, 1983].

For most drugs for which it could be tested (cisplatin, phosphonacetyl-*L*-aspartate, and vinblastine), concentrations in intracerebral tumors appeared to match those in extracerebral tumors; and drug concentrations were generally high for several millimeters beyond the grossly visible edge of the tumor. However, the concentration of etoposide attainable in intracerebral tumors appears to be somewhat lower than the concentrations attained in extracerebral tumors [68]. In addition, concentrations of a given drug may be low in intracerebral tumors of some patients but high in others, and for methylglyoxyl-bis(guanylhydrazone), pentamethylmelamine, and teniposide the concentration achieved may be different for glioblastomas than for brain metastases [56, 68, *Stewart,* unpubl. data, 1983].

It has been suggested that resistance of intracerebral tumors to treatment might be attributed to poor penetration of drug into the zone of brain adjacent to tumor, where tumor cells are actively proliferating [5]. Although this might explain our inability to cure intracerebral tumors with chemotherapy, it should not explain the apparent difficulty in shrinking the main bulk of the tumor.

In brief, the ability of a drug to penetrate the intact BBB is not a prerequisite for that drug's activity against intracerebral tumors, although it may influence the success of treatment in individual patients whose tumor has caused relatively little BBB disruption or when the drug being used has a very steep dose-response relationship. Nevertheless, because the BBB also may have very little effect with some drugs and some tumors, I strongly recommend disregarding a drug's pharmacology in *normal* brain and CSF when initially considering it for investigation in the treatment of intracerebral tumors, whether by means of intravenous or intracarotid infusion.

Techniques of Drug Delivery

For intracarotid infusions, one may use a direct arterial percutaneous puncture technique, insertion of a catheter through the wall of the artery, or a transfemoral, fluoroscopically guided catheter insertion. Arterial puncture is generally used only for short infusions because there is a danger of causing damage to the vessel [37]. Percutaneous catheterization is the method most frequently used for prolonged infusions, but arterial thrombosis and catheter infection may occur, particularly when repeated cannulations are performed. Catheter-associated thrombosis is frequently asymptomatic. In addition, the catheter may become displaced such that drug is infused into perivascular tissues, with resultant compression of vessels and airway and potential necrosis of local tissue. Transient hemiparesis from cannulization-induced vasospasm has also been noted [37, 44, 51, 79]. The superior thyroid artery and superficial temporal artery have been cannulated by some investigators for treatment of brain or head and neck tumors [37]. The transfemoral approach is now used frequently for short-term, intermittent infusions [16, 37, 69] and it is the technique that we prefer. Injections by hand, techniques utilizing gravity, and a variety of pumps have been used to deliver the drug [37, 44, 51, 79].

In addition to the infusion techniques, a variety of perfusion methods have been developed that are intended to decrease systemic concentrations of drug for perfusion. The carotid arteries and jugular veins are cannulated, sometimes concurrently with the use of a neck tourniquet and an epidural balloon to prevent the escape of blood through collateral veins [37, 42]. Such perfusion methods have largely been abandoned, however, because they involve significant technical difficulties.

Results of Treatment

Alkylating Agents

A variety of studies have been performed with nitrogen mustard. In one investigation, intracarotid administration of the drug produced death from marked cerebral edema in 2 of 3 patients [18]. In another, an attempt was made to decrease toxicity by using prolonged infusions of nitrogen mustard (although the instability of the drug limits the usefulness of prolonged administration) together with concurrent intravenous administration of sodium thiosulfate, on the theory that the thiosulfate would bind and inactivate any drug entering the systemic circulation: 4 patients in this series still died from drug-induced cerebral edema [58]. A median survival of 3.8 months was noted in a series of 22 patients with glioblastoma [50, 53]; no therapeutic effect of nitrogen mustard was noted in any of 6 patients with brain metastases [51]. Complications related to toxicity of the drug included ipsilateral conjunctival and scalp erythema, alopecia, and ophthalmoplegia. *Mahaley and Woodhall* [42] used nitrogen mustard perfusion in 2 patients, of whom 1 died from causes related to the treatment.

The alkylating agent melphalan has also undergone limited trials [37]. The rapidly degrading alkylating agent P-(2,3-di(2-bromoethylthio)-n-propyloxy)benzoic acid caused hemiparesis in 2 patients receiving intracarotid infusions for head and neck carcinomas [45]. *Mahaley and Woodhall* [42] have used cyclophosphamide, a drug that requires hepatic activation, for patients with primary brain tumor: 9 patients were treated using cyclophosphamide perfusion and 5 by infusion of the drug; the median survival was 4.5 months. 11 glioblastoma patients treated by carotid infusion of the alkylating agent, N,N′-bis(2(2-chloroethylthio)-propionyl)-1,2-ethylenediammine (S112), had a median survival of 7 months [42]. 13 (52%) of 25 patients with primary brain tumor who were treated with intracarotid triethylene thiophosphoramide (thio-TEPA) [5] responded to this therapy.

Studies in dogs have shown that intracarotid AZQ (2–4 mg/kg) produced minimal toxicity [25]. Recently, clinical studies involving the intracarotid administration of AZQ have been initiated [*Madajewicz*, pers. commun., 1982].

Antimetabolites

Intracarotid MTX resulted in a median survival of 8–11 months [44]. Responses were noted in 4 (33%) of 12 evaluable glioblastoma patients and in 3 of 7 patients with other types of primary brain tumors. Death from neurotoxicity has been reported [23], although intracarotid mannitol given before the administration of MTX resulted in no permanent neurological toxicity in the first 6 patients treated [46]. A single patient treated with intra-arterial 5-FU (650 mg) for a meningosarcoma did not respond to the therapy [74].

Plant Alkaloids

Intracarotid vincristine (1–2 mg daily; total dosage, 10–15 mg) has resulted in a median survival of more than 12 months from the onset of symptoms in 51 glioblastoma patients [51]. Antitumor responses were noted in 7 of 12 evaluable patients [52]. The manifestations of dose-limiting toxicity consisted of ipsilateral conjunctival and forehead erythema, transient ipsilateral ophthalmoplegia, and ileus [51]. Prior irradiation increased the severity of local toxicity.

7 (37%) of 19 patients with primary brain tumors treated by *Wilson* [79] with intracarotid or vertebral artery infusions of vinblastine (0.4–0.5 mg/kg/day) over 3–4 days demonstrated antitumor response, including 6 (46%) of 13 patients who had glioblastomas. The complications resembled those seen with vincristine. *Mealey* [43] noted 2 responses in 10 patients with astrocytoma who were treated with vinblastine (0.5 mg/kg), and *Dean* et al. [12] noted 12 responders among 19 patients.

7 of 10 patients with brain metastases from bronchogenic adenocarcinoma responded to intra-arterial teniposide, including 2 patients who had complete remission [39]. The dosage of 100 mg/m^2 was divided between carotid and vertebral arteries that supplied the areas of metastases. 3 patients experienced transient drowsiness and 1 patient had a seizure. The related epipodophyllotoxin, etoposide, created minimal toxicity in dogs that were given 2.5 mg/kg by intracarotid infusion [58]. Clinical trials with this drug have been initiated recently [*Feun,* pers. commun., 1982].

Cisplatin

We conducted a phase I study of intracarotid cisplatin in patients with intracerebral tumors whose tumor had progressed after irradiation [69]. Cisplatin (60–100 mg/m^2) in 175–250 ml of 0.45% saline with 1,000 units of heparin was infused over 1 h using a transfemoral approach. All patients received concomitant heparin, mannitol, and hydration intravenously. Of 11 patients treated, 7 also received dexamethasone (50 mg) intravenously twice daily for 2 days. Gastrointestinal toxicity was moderate. Neurological toxicity developed in 3 patients treated at a dosage level of 75–100 mg/m^2; 2 of these 3 patients also developed permanent ipsilateral blindness. 1 additional patient developed bilateral decreased visual and auditory acuity 2 weeks after his 6th treatment. 2 patients had orbital pain. Of 10 evaluable patients, 6 had objective responses, as determined by computerized tomography scanning, including 2 of 4 patients with melanoma, 2 of 4 patients with glioblastoma, and 2 other patients, 1 with small cell carcinoma of the lung and 1 with meningeal sarcoma. 2 patients stabilized for 3 and 4 months, respectively.

This work was continued in a phase II study using cisplatin (60 mg/m^2), in which 11 additional patients with glioblastoma were treated [16].

Overall, of 15 glioblastoma patients treated in the 2 studies, 11 were evaluable for response. 5 experienced objective improvement, 4 stabilized, and only 2 failed. In addition, 1 patient with a germ cell tumor was treated during phase II of the study and improved. The phase II study confirmed that retinal and neurological toxicity develops infrequently in patients receiving doses below 75 mg/m^2.

Madajewicz, whose group has also noted responses to intracarotid cisplatin [38], is currently completing studies of melanoma brain metastases [*Madajewicz,* pers. commun., 1982]. *Lehane* et al. [34] have found a cisplatin dose level of 100 mg/m^2 to be tolerable, provided that the drug is filtered through a 2-μm filter before the intracarotid infusion.

Nitrosoureas

Intracarotid BCNU was found to produce neurological dysfunction related to vasculitis in dogs [13]. Less toxicity was noted in monkeys [10]. In patients, the dosage of BCNU (100 mg/m^2) was divided arbitrarily between different carotid and vertebral arteries, depending on the distribution of brain metastases [40]. Most patients developed episcleral vasodilation and periorbital arthralgia that was controlled by topical applications of 0.5% tetracaine hydrochloride ophthalmic solution. Vertebral artery infusion produced occipital and nuchal pain that was decreased by prolonging the duration of the infusion. Some patients developed focal seizures, transient confusion, or mild nausea. Responses occurred in 10 of 25 evaluable patients who had lung cancer, but not in 10 patients with melanoma. Responses occurred also in each of 4 patients with glioblastoma who had had no prior radiation therapy and were treated with intracarotid BCNU (200–300 mg/m^2), and 1 of 2 patients who had prior radiation therapy stabilized [21]; 4 of these 6 patients developed ipsilateral blindness. Additional studies currently under way combine BCNU (600 mg/m^2) given by intracarotid infusion with autologous bone marrow rescue [*Hochberg,* pers. commun., 1982].

Antibiotics

Intracarotid mithramycin (25 μg/kg) given for 20 days did not produce toxicity in 2 patients treated [51]; the patients survived 2 and 4 months, respectively. Intracarotid neocarzinostatin resulted in response in 10 of 12 patients with brain tumors [41]. Intracarotid doxorubicin in a dosage of 1 mg/kg did not produce neurological toxicity in dogs, although marked neurological toxicity was noted with lower doses if an intracarotid dosage of mannitol preceded the doxorubicin [47]. The related anthracycline, quelamycin, produced severe cerebral edema with permanent neurological damage and death in some patients with glioblastoma who were treated by intracarotid infusion [6]; nevertheless, antitumor efficacy, including complete remissions in 3 patients, was noted with this drug. The concurrent administration of high-dose dexamethasone, forced diuresis, and the use of low weekly doses and prolonged infusions are now being investigated in an attempt to decrease the toxicity of quelamycin.

We have used intracarotid mitomycin-C (15 mg/m^2 in 250 ml 5% dextrose-water solution) in treating 2 patients with glioblastoma and 2 patients with brain metastases [*Stewart,* unpubl. data, 1983]. A patient with brain metastases from testicular carcinoma has had marked shrinkage of the tumor that continues 9 months after the initiation of therapy. Decreased visual acuity was noted in the ipsilateral eye after his 4th treatment, and thrombocytopenia has been noted with each course. Treatments are repeated at 9-week intervals. The other 3 patients treated with intracarotid mitomycin C had no observed toxicity or response. Further clinical trials with this drug are planned, although it is known to have a propensity to cause arterial damage.

Other Agents

External carotid artery infusion of the hormone Premarin® led to occlusion of tumor vessels and necrosis of 3 meningiomas treated [70]; the major form of toxicity was parotitis. 5-Bromo-2′-deoxyuridine, a radiosensitizer that is incorporated into the DNA of rapidly growing cells, has been administered by intracarotid infusion [37]. Intracarotid injection of radioactive yttrium-90 microspheres resulted in improvement in approximately half of 15 patients with glioblastoma and 10 with brain metastases [60]; infusion of radioactive material into the ophthalmic artery was prevented by compressing the eye while the microspheres were being administered.

Intracarotid Combination Chemotherapy

Combination chemotherapy has been used only occasionally in intraarterial treatment. *Mealey* et al. [44] observed no major responses in 7 patients treated concurrently with mithramycin and vincristine. Hemiplegia, ipsilateral oculomotor palsy, and Horner's syndrome developed in a few patients.

26 patients with malignant gliomas received intracarotid BCNU

(100 mg/m^2) combined with vincristine (1 mg/m^2) intravenously on the same day followed by oral doses of procarbazine (100 mg/m^2) daily for 14 days [80]. Of these 26 patients, 22 were evaluable for response; 2 of 8 who had previously had radiation therapy showed response to this combination, whereas 10 of 14 who had not undergone irradiation responded. The responses lasted 32+ to 81+ weeks. Focal seizures, transient orbital pain, periorbital and scleral erythema, conjunctival suffusion, lacrimation and pulmonary emboli were frequent complications.

Recently, studies have also been initiated to evaluate the efficacy of the intracarotid infusion of BCNU in combination with either cisplatin [*Feun,* pers. commun., 1982] or the epipodophyllotoxin, VM-26 [*Madajewicz,* pers. commun., 1982]. We have treated 18 patients with the 3-drug combination of BCNU (100 mg/m^2), cisplatin (40–60 mg/m^2) and VM-26 (100–175 mg/m^2) given by intracarotid infusion [*Stewart,* unpubl. data, 1983]. All patients were given dexamethasone (50 mg) intravenously once before and once after their treatment. Intravenous mannitol (50 g) was also given to patients receiving more than 50 mg/m^2 of cisplatin. Transient orbital discomfort, nausea, and myelosuppression developed in most patients. 2 patients also developed transient hemiparesis and aphasia after the treatment, and another had intermittent myoclonic reactions while the VM-26 was being infused. 2 patients have experienced ipsilateral retinal toxicity with decreased visual acuity. Most patients became quite somnolent within 5 min of initiation of BCNU and VM-26 infusions, but became fully alert again within 5 min after the infusion was terminated. The only patient to receive a vertebral artery infusion of the drugs had rapidly reversible respiratory depression during the BCNU and VM-26 infusions. It is too early to evaluate the response status in 4 patients, and 1 patient was not evaluable because he died within 2 weeks of treatment. 4 of 6 evaluable patients with glioblastoma and 4 of 8 evaluable patients with brain metastases responded, despite all but 2 having had tumor recurrence after cranial irradiation. 8 of the 14 evaluable patients had previously received 1 or more of the following: BCNU, cisplatin, and VM-26 intravenously. Further studies with this combination of drugs are under way.

Prospects and Limitations

Like many aspects of cancer therapy, intracarotid infusion of chemotherapeutic agents for the treatment of intracerebral neoplasms is a 'half-way technology'. Once highly effective drugs or drug combinations are developed, it is likely that it will be possible to cure brain tumors by means of systemic chemotherapeutic administration. In the meanwhile, it is worth pursuing studies with drugs given by intracarotid infusion, as this method offers at least the potential for achieving palliation.

The main difficulties with this form of therapy currently relate to the necessity for either repeated arterial catheterizations or permanent catheterization and the attendant risks of carotid artery thrombosis and infection. Recently, a totally implantable infusion pump, Infusaid®, has become available that has been tested in only a small number of patients with brain tumors [55]. In that group, a catheter from the pump was inserted into the external carotid artery and threaded down to the carotid bifurcation. The external carotid artery was ligated distally, and the superior thyroid artery was ligated also, so that all infused chemotherapeutic agent would enter the internal carotid artery. Of the 6 patients studied, none developed thromboembolic complications, although retinal toxicity was noted. To prevent retinal toxicity, other types of catheterization procedures might be feasible. For example, it is possible that the superficial temporal artery could be implanted end-to-side into a tumor vessel above the ophthalmic artery. If a catheter were used that exactly filled the lumen of the implanted artery, if the catheter did not project out from the implanted artery, and if a constant infusion were maintained by a reliable pump, then it is possible that thromboembolic complications could be avoided. This and similar approaches deserve at least preliminary feasibility studies in animals.

Initial studies are also under way that involve superselective percutaneous transfemoral catheterization of the carotid artery, such that the tip of the catheter is placed above the ophthalmic artery. Methods such as placing pressure on the eye intermittently during intracarotid infusion in order to decrease the ophthalmic artery flow [60] also deserve further study as a possible means for decreasing periorbital pain and retinal toxicity.

Frequently, local toxicity – not systemic toxicity – limits the drug dosage that can be given by intracarotid infusion. Further studies [78] in which systemic chemotherapy is added to intracarotid chemotherapy are warranted. While potentially there are great risks involved in blocking the blood supply to the brain, transient proximal occlusion of the artery with miniature balloon catheters during infusion [54] could provide markedly enhanced local concentrations of drug and better mixing of drug in the bloodstream. In addition, judging from the experience with the injection of yttrium microspheres [60], microspherule injection performed concurrently with carotid artery infusions might be tolerated by the patient; if so, this technique could be used to augment local uptake of drug in tissue, as has been observed when intrahepatic infusions are used.

The efficacy of therapy with intracarotid infusions of extremely high doses of chemotherapy coupled with marrow-sparing autologous bone marrow infusions is now under investigation [*Hochberg,* pers. commun., 1982]. The initial studies warrant continued investigation.

The proposals described here are only a few examples of investigations that could prove fruitful in the future. Carotid artery infusion remains fraught with difficulties, and this technique is still of only limited clinical benefit. Nevertheless, with continued research, it could prove to be an important therapeutic modality for the management of intracerebral tumors.

Intrathecal Route

The degree of penetration of a drug into brain after intrathecal injection varies, depending on the agent used [4]. When BCNU or thio-TEPA is administered intrathecally, the drug attains only low concentrations in brain because it readily enters blood vessels in the initial few millimeters of the brain and is washed out into the systemic circulation. Cytosine arabinoside also attains only low concentrations in brain after intrathecal injection because it is metabolized rapidly by brain tissue. On the other hand, hydroxyurea and MTX may penetrate quite deeply into brain tissue after intrathecal injection, although they do so quite slowly. There is some data to indicate that concentrations of MTX in both human and animal brain tumors may be higher when the drug is administered intrathecally than when it is given intravenously [49, 75], and that the concentration of drug is higher in tumor than in surrounding brain [75]. Intraventricular administration of MTX results in higher concentrations in brain than does injection into the lumbar sac [80]. The injection of MTX entrapped in cholesterol liposomes appears to result in higher average MTX concentrations in brain than does injection of free MTX, although the entrapped MTX may not penetrate as far into deep areas of the brain as the free MTX does [31].

MTX is the only drug that has been used in studies of intrathecal chemotherapy for solid intracerebral tumors. Subjective and objective improvement in patients treated by this approach has been documented, although in most cases, the duration of improvement was brief [11, 48]. When a prolonged, continuous intraventricular infusion of MTX was given using an Infusaid® pump, some patients developed neurological toxicity [11]. Long-term remissions of primary lymphoma of the central nervous system have been noted following intrathecal treatment with MTX [26].

Intratumoral Route

Pharmacological studies in animals suggest that the intratumoral injection of MTX results in rapid spread of the drug through the tumor, although the distribution is uneven [72]. Nonetheless, intratumoral injections of MTX did not appear to be effective either against a murine ependymoblastoma model [73] or against human glioma [19]. Intratumoral CCNU was somewhat effective against the murine ependymoblastoma line [71], whereas intratumoral BCNU appeared to be of limited benefit against human glioma [20]. Intratumoral bleomycin injections proved superior to bleomycin given intravenously in a rat gliosarcoma model [76] and pharmacology studies suggest that systemic exposure to bleomycin is less when the drug is injected intratumorally than when it is administered intravenously [2]. Studies of other agents injected intratumorally are also in progress. Overall, intrathecal and intratumoral injection of drugs have not proved very effective in the treatment of human intracerebral tumors.

References

1 Allen, J.C.; Hosegawa, H.; Metha, B.M.; Shapiro, W.R.; Posner, J.B.: Influence of intracarotid mannitol on intracerebral methotrexate (MTX) concentrations surrounding experimental brain tumors. Proc. Am. Ass. Cancer Res. *20:* 286 (1979).

2 Bier, J.; Benders, P.; Wenzel, M.; Bitter, K.: Kinetics of ^{57}Co-bleomycin in mice after intravenous, subcutaneous, and intratumoral injection. Cancer *44:* 1194–1200 (1979).

3 Blasberg, R.G.: Pharmacodynamics and the blood-brain barrier. Natn. Cancer Inst. Monogr. *46:* 19–27 (1977).

4 Blasberg, R.G.; Patlak, C.; Fenstermacher, J.D.: Intrathecal chemotherapy in brain tissue profiles after ventriculocisternal perfusion. J. Pharmac. exp. Ther. *195:* 73–83 (1975).

5 Broder, L.E.; Rall, D.P.: Chemotherapy of brain tumors. Prog. exp. Tumor Res., vol. 17, pp. 373–399 (Karger, Basel 1972).

6 Brugarolas, A.; Garcia de Sola, R.; Bravo, G.; Lahoz, C.: Intraarterial quelamycin (QLM, triferric adriamycin, NSC 267703) in the treatment of glioblastoma. Proc. Am. Soc. clin. Oncol. *21:* 346 (1980).

7 Calvo, D.B., III; Patt, Y.Z.; Wallace, S.; Chuang, V.P.; Benjamin, R.S.; Pritchard, J.D.; Hersh, E.M.; Bodey, G.P., Sr., Mavligit, G.M.: Phase I–II trial of percutaneous intra-arterial cis-diamminedichloroplatinum (II) for regionally confined malignancy. Cancer *45:* 1278–1283 (1980).

8 Campbell, T.N.; Howell, S.B.; Wung, W.E.; Bookstein, J.; Pfeifle, C.E.: Pharmacokinetics of intra-arterial (IA) cisplatin (DDP). Proc. Am. Soc. clin. Oncol. *1:* 27 (1982).

9 Chen, H.-S.G.; Gross, J.F.: Intra-arterial infusion of anticancer drugs: theoretic aspects of drug delivery and review of responses. Cancer Treat. Rep. *64:* 31–40 (1980).

10 Crafts, D.C.; Levin, V.A.; Nielsen, S.: Intracarotid BCNU (NSC 409962). A toxicity study in six rhesus monkeys. Cancer Treat. Rep. *60:* 541–545 (1976).

11 Dakhil, S.; Ensmeinger, W.; Kindt, G.; Riederhuber, J.; Chandler, W.; Greenberg, H.; Wheeler, R.: Implanted system for intraventricular drug infusion in central nervous system tumors. Cancer Treat. Rep. *65:* 401–411 (1981).

12 Dean, M.R.F.; Newton, K.A.; Swann, G.F.: Percutaneous intra-arterial chemotherapy in the treatment of intracranial neoplasms. A review of 36 cases. Br. J. Radiol. *40:* 828–833 (1967).

13 De Wys, W.D.; Fowler, E.H.: Report of vasculitis and blindness after intracarotid injection of 1,3-bis-(2-chloroethyl)-1-nitrosourea (BCNU, NSC 409962) in dogs. Cancer Chemother. Rep. *57:* 33–40 (1973).

14 Dorr, R.T.; Fritz, W.L.: Cancer chemotherapy handbook, p.269 (Elsevier/North-Holland, New York 1980).

15 Fenstermacher, J.D.; Cowles, A.K.: Theoretic limitations of intracarotid infusions in brain tumor chemotherapy. Cancer Treat. Rep. *61:* 519–526 (1977).

16 Feun, L.; Wallace, S.; Stewart, D.; Leavens, M.; Savaraj, N.; Burgess, M.A.; Young, S.; Benjamin, R.S.: Phase I–II trial of intracarotid (IC) cis-diamminedichloroplatinum (DDP) in patients with intracerebral tumors (ICT). Proc. Am. Soc. clin. Oncol. *1:* 197 (1982).

17 Frankel, S.A.; German, W.J.: Glioblastoma multiforme: review of 219 cases with regard to natural history, pathology, diagnostic methods and treatment. J. Neurosurg. *15:* 489–503 (1958).

18 French, J.D.; West, P.M.; Van Amerongen, F.K.; Magoun, H.W.: Effects of intracarotid administration of nitrogen mustard on normal brain and brain tumors. J. Neurosurg. *9:* 378–389 (1952).

19 Garfield, J.; Dayan, A.D.: Postoperative intracavitary chemotherapy of malignant gliomas. J. Neurosurg. *39:* 315–322 (1973).

20 Garfield, J.; Dayan, A.D.; Weller, R.O.: Postoperative intracavitary chemotherapy of malignant supratentorial astrocytomas using BCNU. Clin. Oncol. *1:* 213–222 (1975).

21 Greenberg, H.S.; Ensminger, W.D.; Seeger, J.F.; Kindt, G.W.; Chandler, W.F.; Doan, K.; Dakhil, S.R.: Intra-arterial BCNU chemotherapy for the treatment of malignant gliomas of the central nervous system: a preliminary report. Cancer Treat. Rep. *65:* 803–810 (1981).

22 Greenfield, A.J.; Athasoulis, C.A.; Waltman, A.C.; Le Moure, E.R.: Transcatheter

embolization: prevention of embolic reflux with balloon catheters. Am. J. Roentg. *131:* 651–655 (1978).
23 Greenhouse, A.J.; Neubuerger, J.T.; Bowerman, D.L.: Brain damage after intracarotid infusion of methotrexate. Archs Neurol. *11:* 618–625 (1964).
24 Groothius, D.R.; Fischer, J.M.; Vick, N.A.; Bigner, D.D.: Comparative permeability of different glioma models to horseradish peroxidase. Cancer Treat. Rep. *65:* suppl. 2, pp. 13–18 (1981).
25 Guo, Z.G.; Lu, K.; Savaraj, N.; Feun, L.G.; Huang, M.T.; Loo, T.L.: Pharmacology fate of 2,5-diaziridinyl-3,6-biscarboethoxyamino-1,4-benzoquinone (AZQ, NSC 182986) by intracarotid (IC) or intravenous (IV) administration in beagles. Proc. Am. Ass. Cancer Res. *23:* 218 (1982).
26 Herbst, K.D.; Corder, M.P.; Justice, G.R.: Successful therapy with methotrexate of a multicentric mixed lymphoma of the central nervous system. Cancer *38:* 1476–1478 (1976).
27 Kaplan, W.D.; d'Orsi, C.J.; Ensminger, W.D.; Smith, E.H.; Levin, D.C.: Intra-arterial radionuclide infusion: a new technique to assess chemotherapy perfusion patterns. Cancer Treat. Rep. *62:* 699–703 (1978).
28 Karakousis, C.P.; Sharma, S.D.; Kanter, P.: Regional chemotherapy to the canine liver with 5-fluorouracil (5FU) and adriamycin (AM). Proc. Am. Ass. Cancer Res. *23:* 211 (1982).
29 Kato, T.; Nemoto, R.; Mori, H.; Takahashi, M.; Tamakawa, Y.; Harada, M.: Arterial chemoembolization with microencapsulated anticancer drugs. J. Am. med. Ass. *245:* 1123–1127 (1981).
30 Khan, A.B.; d'Souza, B.; Wharam, M.; Sinks, L.; Woo, S.; McCullough, D.; Leventhel, B.G.: Cisplatinum in recurrent brain tumors. Proc. Am. Soc. clin. Oncol. *22:* 402 (1981).
31 Kimelberg, H.K.; Tracy, T.F.; Watson, R.E.; Kung, D.; Reiss, F.L.; Bourke, R.S.: Distribution of free and liposome-entrapped (^{3}H)-methotrexate in the central nervous system after intracerebral injection in a primate. Cancer Res. *38:* 706–712 (1978).
32 Kolaric, K.; Roth, A.; Pavelic, Z.: Phase II clinical trials of cis-dichlorodiammine platinum in metastatic brain tumors. Proc. Am. Soc. clin. Oncol. *1:* 183 (1981).
33 Legha, S.S.; Benjamin, R.S.; MacKay, B.; Ewer, M.; Wallace, S.; Valdivieso, M.; Rasmussen, R.T.; Blumenschein, G.R.; Freireich, E.J.: Reduction of doxorubicin cardiotoxicity by prolonged continuous intravenous infusion. Ann. intern. Med. *96:* 133–139 (1982).
34 Lehane, D.; Ehni, G.; Moiel, R.; Bryon, R.; Horowitz, B.; De Santos, L.; Aldama-Luebert, A.; Zublir, M.; Rudolph, L.; Lane, M.: Adjuvant therapy for malignant gliomas (MG) using intra-arterial (IA) cis-platinum (CP), radiation therapy (RT) and CCNU. Proc. Am. Soc. clin. Oncol. *2:* 236 (1983).
35 Lehane, D.E.; Zubler, M.A.; Lane, M.; Smith, F.E.: Intra-arterial cisplatin in metastatic or recurrent colon cancer. Proc. Am. Soc. clin. Oncol. *1:* 94 (1982).
36 Levin, V.A.; Kabra, P.M.; Freeman-Dove, M.A.: Pharmacokinetics of intracarotid artery ^{14}C-BCNU in the squirrel monkey. Neurosurgery *48:* 587–593 (1978).
37 Levin, V.A.; Wilson, C.B.: Pharmacological considerations in brain tumor chemotherapy; in Fewer, Wilson, Levin, Brain tumor chemotherapy, pp. 56–60 (Thomas, Springfield 1976).
38 Madajewicz, S.; Kanter, P.; West, C.; Bhargava, A.; Prajapati, R.; Caracandas, J.;

Avellanosa, A.; Fitzpatrick, J.: Plasma, spinal fluid and organ distribution of cisplatinum (DDP) following intravenous (IV) and intracarotid (IC) infusion. Proc. Am. Ass. Cancer Res. *22:* 176 (1981).

39 Madajewicz, S.; West, C.R.; Park, H.C.; Avellanosa, A.M.; Caracandas, J.E.; Takita, H.; Vincent, R.G.; Jennings, E.: Phase II study of intra-arterial 4′-0-demethyl-1-0-(4,6-0-2-thenylidene-B-*D*-glucopyranosyl) epipodophyllotoxin (VM-26) therapy of metastatic bronchogenic adenocarcinoma to the brain. Proc. Am. Soc. clin. Oncol. *22:* 502 (1981).

40 Madajewicz, S.; West, C.R.; Park, H.Y.; Ghoorah, J.; Avellanosa, A.M.; Takita, H.; Karakousis, C.; Vincent, R.; Caracandas, J.; Jennings, E.: Phase II study intra-arterial BCNU therapy for metastatic brain tumors. Cancer *47:* 653–657 (1981).

41 Maeda, H.; Sano, Y.; Takishita, J.; Iwai, Z.; Kosaka, H.; Marubayashi, T.; Matsukado, Y.: A pharmacokinetic simulation model for chemotherapy of brain tumor with an antitumor protein antibiotic, neocarzinostatin. Cancer Chemother. Pharmacol. *5:* 243–249 (1981).

42 Mahaley, M.S., Jr.; Woodhall, B.: Regional chemotherapeutic perfusion and infusion of brain and face tumors. Ann. Surg. *166:* 266–277 (1967).

43 Mealey, J., Jr.: Treatment of malignant cerebral astrocytomas by intra-arterial infusion of vinblastine. Cancer Chemother. Rep. *20:* 121–126 (1962).

44 Mealey, J., Jr.; Chen, T.T.; Pedlow, E.: Brain tumor chemotherapy with mithramycin and vincristine. Cancer *26:* 360–367 (1970).

45 Meyza, J.; Cobb, L.M.: The clinical trial of an alkylating agent with a short half-life designed for intra-arterial chemotherapy. Cancer *27:* 369–373 (1971).

46 Neuwelt, E.A.; Frenkel, E.P.; Diehl, J.T.; Hill, S.A.: Initial clinical studies of chemotherapy delivery to brain tumor patients after osmotic blood-brain barrier modification. Proc. Am. Soc. clin. Oncol. *22:* 351 (1981).

47 Neuwelt, E.A.; Pagel, M.; Barnett, P.; Glassberg, M.; Frenkel, E.P.: Pharmacology and toxicity of intracarotid adriamycin administration following osmotic blood-brain barrier modification. Cancer Res. *41:* 4466–4470 (1981).

48 Newton, W.A., Jr.; Sayers, M.P.; Samuels, L.D.: Intrathecal methotrexate (NSC-740) therapy for brain tumors in children. Cancer Chemother. Rep. *52:* 257–261 (1968).

49 Ojima, Y.; Sullivan, R.D.: Pharmacology of methotrexate in the human central nervous system. Surgery Gynec. Obstet. *125:* 1035–1040 (1967).

50 Owens, G.: Infusion chemotherapy. N. Y. med. J. *66:* 3026–3029 (1966).

51 Owens, G.: Intra-arterial chemotherapy of primary brain tumors. Ann. N. Y. Acad. Sci. *159:* 603–607 (1969).

52 Owens, G.; Jarid, R.; Belmusto, L.; Bender, M.; Blau, M.: Intra-arterial vincristine therapy of primary gliomas. Cancer *18:* 756–760 (1965).

53 Owens, G.; Jarid, R.; Tallon, M.; Stepanian, G.; Belmusto, L.: Arterial infusion chemotherapy of primary gliomas. J. Am. med. Ass. *186:* 802–803 (1963).

54 Pevsner, P.H.; Doppman, J.; Lipper, N.H.; Holton, J.C.; Young, H.F.; Dillard, D.C.: Therapeutic cerebrovascular occlusion and perfusion of angiomas and neoplasms. Neuroradiology *15:* 53 (1978).

55 Phillips, T.W.; Chandler, W.F.; Kindt, G.W.; Ensminger, W.D.; Greenberg, H.S.; Seezer, J.F.; Doan, K.M.; Gywes, J.W.: New implantable continuous administration and bolus dose intracarotid drug delivery system for the treatment of malignant gliomas. Neurosurgery *11:* 213–218 (1982).

56 Rosenblum, M.; Stewart, D.J.; Yap, B.S.; Leavens, M.; Benjamin, R.S.; Loo, T.L.: Penetration of methylglyoxal bis (guanylhydrazone) into intracerebral tumors in humans. Cancer Res. *41:* 459–462 (1981).

57 Rosner, D.; Nemoto, T.: Blood-brain barrier a myth? Chemotherapy of central nervous system (CNS) metastases in breast cancer. A pilot study. Proc. Am. Ass. Cancer Res. *24:* 140 (1983).

58 Savaraj, N.; Lu, K.; Rauston, G.L.; Feun, L.G.; Loo, T.L.: Comparison of CNS penetration, tissue distribution and pharmacology of 4′-demethyl epipodophyllotoxin 9-(4,5-0-ethylidene-B-*D*-glucopyranoside) (VP-16, NSC-141540). Intracarotid (IC) vs. intravenous administration (IV) in dogs. Proc. Am. Ass. Cancer Res. *23:* 201 (1982).

59 Savaraj, N.; Lu, K.; Stewart, D.J.; Leavens, M.; Bedekian, A.; Feun, L.; Loo, T.L.: Tissue distribution and intracerebral tumor penetration of 2,5-diaziridinyl-3,6-biscarboethoxyamino-1,4-benzoquinone (AZQ, NSC 182986) in man. Proc. Am. Soc. clin. Oncol. *22:* 351 (1981).

60 Simon, M.: Intra-arterial irradiation of tumors of the brain and liver. Panminerva med. *13:* 338–339 (1971).

61 Stewart, D.J.; Benjamin, R.S.: Central nervous system dysfunction secondary to chemotherapeutic agents; in Levenson, Neuropsychiatric side effects of drugs in the elderly, pp. 204–205 (Raven Press, New York 1979).

62 Stewart, D.J.; Benjamin, R.S.; Luna, M.; Seifert, W.E.; Loo, T.L.: Human tissue distribution of platinum after cis-diamminedichloroplatinum. Cancer Chemother. Pharmacol. *10:* 51–54 (1982).

63 Stewart, D.; Benjamin, R.S.; Luna, M.; Yeung, K.; Leavens, M.; Feun, L.; Yap, H.Y.; Loo, T.L.: Human tissue vinblastine (V) and cisplatin (CP). Proc. Am. Ass. Cancer Res. *23:* 125 (1982).

64 Stewart, D.J.; Benjamin, R.S.; Zimmerman, S.; Caprioli, R.M.; Wallace, S.; Chuang, V.; Calvo, D., III; Samuels, M.; Bonura, J.; Loo, T.L.: Clinical pharmacology of intra-arterial cis-diamminedichloroplatinum. Cancer Res. *43:* 917–920 (1983).

65 Stewart, D.J.; Benvenuto, A.J.; Leavens, M.; Hall, S.W.; Benjamin, R.S.; Plunkett, W.; McCredie, K.D.; Burgess, M.A.; Loo, T.L.: Penetration of 3-deazauridine into human brain, intracerebral tumor, and cerebrospinal fluid. Cancer Res. *39:* 4119–4122 (1979).

66 Stewart, D.J.; Leavens, M.; Friedman, J.; Benjamin, R.S.; Moore, E.C.; Bodey, G.P.; Valdivieso, M.; Burgess, M.A.; Wiseman, C.; Loo, T.L.: Penetration of N-(phosphonacetyl)-*L*-aspartate into human central nervous system and intracerebral tumor. Cancer Res. *40:* 3163–3166 (1980).

67 Stewart, D.J.; Leavens, M.; Maor, M.; Feun, L.; Luna, M.; Bonura, J.; Caprioli, R.; Loo, T.L.; Benjamin, R.S.: Human central nervous system distribution of cis-diamminedichloroplatinum and use as a radiosensitizer in malignant brain tumors. Cancer Res. *42:* 2474–2479 (1982).

68 Stewart, D.J.; Richard, M.; Hugenholtz, H.; Dennery, J.: VP-16 (VP) and VM-26 (VM) penetration into human brain tumors (BT). Proc. Am. Ass. Cancer Res. *24:* 133 (1983).

69 Stewart, D.J.; Wallace, S.; Feun, L.; Leavens, M.; Young, S.E.; Handel, S.; Mavligit, G.; Benjamin, R.S.: A phase I study of intracarotid artery infusion of cis-diamminedichloroplatinum in patients with recurrent malignant intracerebral tumors. Cancer Res. *42:* 2059–2062 (1982).

70 Suzuki, J.; Komatsu, S.: New embolization method using estrogen for dural arteriovenous malformation and meningioma. Surg. Neurol. *16:* 438–442 (1981).
71 Tator, C.H.: Intraneoplastic injection of CCNU for experimental brain tumor chemotherapy. Surg. Neurol. *7:* 73–77 (1977).
72 Tator, C.H.; Wassenaar, W.: Intraneoplastic injection of methotrexate for experimental brain-tumor chemotherapy. J. Neurosurg. *46:* 165–174 (1977).
73 Tator, C.H.; Wassenaar, W.; Day, A.; So, W.S.: Therapy of an experimental glioma with systemic or intraneoplastic methotrexate or radiation. J. Neurosurg. *46:* 175–184 (1977).
74 Tindel, S.: Intra-arterial chemotherapy for recurrent neoplasms. J. Am. med. Ass. *200:* 913–917 (1967).
75 Ushio, Y.; Hayakawa, T.; Mogami, H.: Uptake of tritiated methotrexate by mouse brain tumors after intravenous or intrathecal administration. J. Neurosurg. *40:* 706–716 (1974).
76 Vats, T.S.; Morantz, R.A.; Wood, G.W.; Tilzer, S.: Study of effectiveness of bleomycin in rat brain tumor model intravenously and intracerebrally. Int. J. Radiat. Oncol. Biol. Phys. *5:* 1527–1529 (1979).
77 Vick, N.A.; Khandekar, J.D.; Bigner, D.D.: Chemotherapy of brain tumors: the 'blood-brain barrier' is not a factor. Archs Neurol. *34:* 523–526 (1977).
78 West, C.R.; Avellanosa, A.M.; Varba, N.R.; Patel, A.: Phase II study of malignant gliomas of the brain treated with intra-arterial BCNU in combination with vincristine and procarbazine. Proc. Am. Soc. clin. Oncol. *21:* 482 (1980).
79 Wilson, C.B.: Chemotherapy of brain tumors by continuous arterial infusion. Surgery *55:* 640–653 (1964).
80 Yen, J.; Reiss, F.L.; Kimelberg, H.K.; Bourke, R.S.: Direct administration of methotrexate into the central nervous system of primates. J. Neurosurg. *48:* 895–902 (1978).
81 Zhengang, G.; Stewart, D.; Lu, K.; Savaraj, N.; Leavens, M.; Feun, L.G.; Benjamin, R.; Loo, T.L.: Tumor penetration of AMSA in man. Proc. Am. Soc. clin. Oncol. *1:* 17 (1982).

D.J. Stewart, MD, The Ontario Cancer Foundation, Ottawa Clinic, General Hospital Division, 501 Smyth Road, Ottawa, Ont. K1G 8L6 (Canada)

Prog. exp. Tumor Res., vol. 28, pp. 51–66 (Karger, Basel 1984)

Therapeutic Potential for Blood-Brain Barrier Modification in Malignant Brain Tumor

Edward A. Neuwelt

Departments of Surgery (Division of Neurosurgery) and Biochemistry, Oregon Health Sciences University, Portland, Oreg., USA

Recent successes with multiagent chemotherapy in the management of a variety of cancers have provided the basis for an important adjustment of focus on the biology and therapy of malignant lesions identified in the central nervous system (CNS). Several types of disseminated systemic malignancies (choriocarcinoma, testicular carcinoma, oat cell carcinoma of lung, and breast cancer in women) managed with chemotherapy have shown significant and durable complete response rates, even when the tumor had metastasized to multiple sites. It is remarkable, therefore, that when these same responsive neoplasms are identified as metastatic foci in the CNS, the effect of the chemotherapy at the CNS sites has been minimal [26]. Similarly, systemic or intrathecal chemotherapy has not had a significant impact on primary CNS tumors.

The unique biologic circumstance of a blood-brain barrier (BBB) is a focal point in the consideration of therapy of tumors in the CNS. A serious proposal relative to the documented ineffectiveness of therapy in such lesions is that the presence of the BBB at the proliferating edge of primary and metastatic CNS tumors is a principal basis for this failure. It is the purpose of this study to consider some of the characteristics of this barrier and to evaluate the efficacy and toxicity of disruption of the barrier before the administration of systemic chemotherapy.

BBB Limits Delivery of Drugs

The BBB is created by tight junctions (zonulae occludentes) between endothelial cells in brain capillaries. These junctions restrict diffusion between endothelial cells; they are found wherever there is physiologic evidence of a barrier and are absent in the few areas that lack a BBB, such as the pineal gland, area postrema, and adenohypophysis [27].

Vick and Bigner [34] have suggested that the 'barrier hypothesis' is irrelevant because earlier studies of the BBB suggested that 'the blood-brain barrier was not a factor in the chemotherapy of brain tumors'. In general, this thesis was based on the observation that the vessels inside metastatic cerebral tumors often have a fenestrated and discontinuous endothelium that is characteristic of the vessels of the primary tumor [10]; similar fenestrated and gap junctions have also been seen in gliomas [10]. Nevertheless, current studies have confirmed that the 'barrier' is far more complex. There may be loss of the BBB in the center of the tumor, but recent studies have provided important evidence that the tumor periphery has very different characteristics. For instance, *Levin* et al. [9] studied uptake of methotrexate (MTX) by intracerebral ependymoblastomas. The MTX level of normal brain was 3–4% of the plasma level; but in the center of the tumor it was 32%, and at the brain-tumor interface 13% of the plasma level. Similarly, *Tator* [33] gave radiolabeled MTX intravenously to mice with intracerebral implants of ependymoblastoma; cells in the central mass were heavily labeled, whereas cells at the periphery and those infiltrating adjacent brain showed only scant labeling. *Shapiro* et al. [32], using ^{14}C-labeled α-aminoisobutyric acid, also reported variability in the degree of BBB disruption in experimental brain tumor; they found a nearly normal BBB in small tumors and adjacent brain. *Groothuis* et al. [6] showed that intravascular peroxidase variably penetrated experimental gliomas in the brain but freely entered subcutaneous tumors. *Walker and Weiss* [35] proposed a 'sink effect' to explain the low concentration of drugs at the tumor periphery; that is, the 'extra' drug that enters the periphery of the tumor rapidly diffuses into drug-free surrounding brain because of 'mild disruption' of the BBB.

That the drug delivery problem is clinically significant was emphasized by *Benjamin* et al. [2], who described progressive increase in the size of brain metastases while other systemic metastases regressed with adriamycin chemotherapy. *Posner* [26] stated, 'drugs that appear to be successful in treating systemic metastases are ineffective against the same tumor when it is in the brain'. It is not known whether these failures are the result of only a partially impaired barrier at the periphery of the tumor, of the 'sink effect', or both.

Two approaches have been taken to overcome these problems in the chemotherapy of neoplasms in the CNS. First, as MTX normally penetrates the brain, albeit poorly, attempts to overcome the resistance of the BBB have involved the use of very high systemic doses of MTX [31]. Unfortunately, these resulted in only marginal therapeutic levels of MTX in brain. A second approach has involved direct instillation of drugs into neural spaces (e.g., the ventricular, subarachnoid). *Kimelberg* et al. [8] gave MTX by direct intraventricular infusion but failed to obtain therapeutic drug levels in the white matter deep to the cortex or the ependyma. With intrathecal or intraventricular chemotherapy, another problem is that the extent of drug dispersal cannot be determined or controlled, especially when there is blockage of cerebrospinal fluid (CSF) circulation (in spinal cord or tentorial lesions).

Regulation of Drug Entry into the Brain

The two important factors determining entry of the drug from blood into the brain are its molecular weight and lipid solubility. The BBB normally prevents passage of ionized water-soluble drugs with a molecular weight of more than 180 daltons [5]. Most of the currently effective chemotherapeutic agents have a molecular weight of between 200 and 1,200 daltons (MTX, 455; daunorubicin, 544; cyclophosphamide, 261). Thus, on a molecular-weight basis, the passage of many chemotherapeutic agents would be impeded by the BBB. An even more important factor is lipid solubility. There is a linear correlation between the octanol/water coefficient of chemotherapeutic agents and cerebrovascular permeability. MTX, which has a pH of 4.7 and is 99.8% ionized at a blood pH of 7.4, is lipid-insoluble; the normal CSF/plasma ratio is only 0.02 [27]. Entry of macromolecular water-soluble nutrients and drugs when the BBB is intact is also regulated by a number of transport mechanisms. The cerebrovascular endothelium can transport, by facilitated, stereospecific, saturable mechanisms, substances that are involved in brain metabolism.

Reversible Disruption of the BBB

To achieve effective therapy, the BBB disruption must be reversible. Only a few techniques for this purpose are known. Hypercarbia disrupts the BBB reversibly but inconsistently [27]. BBB disruption by intracarotid infusion of hypertonic solutions is the most thoroughly evaluated and most effective method for achieving reversible BBB disruption [30].

Reversible osmotic BBB disruption is a threshold event with regard to either osmolality or duration of infusion [30]. Osmotic barrier disruption is essentially an all-or-nothing phenomenon. For instance, in a study using arabinose in the rat, the threshold was 1.6 osm and the minimum infusion time was 20–30 s [30]. Similar thresholds have been observed in rabbit, rat, and primates. After disruption, the barrier remains open for less than 1 h [12, 30].

The importance of the rate of carotid infusion was demonstrated by direct observation of the cerebral cortical surface when infusates were just above the threshold of osmolality [12, 30]. Although this rate measurement is technically complex, it can be obtained by determining the rate of infusion which, by visual inspection, completely displaced blood from the cortical surface of the ipsilateral hemisphere. This rate is critical, as dilution of the infusate by blood from either the ipsilateral carotid artery or from collateral flow via the circle of Willis results in a subthreshold osmolality. Unfortunately, infusion rates higher than necessary result in intravascular hypertension, which can disrupt the barrier irreversibly.

Osmotic disruption of the BBB can be observed ultrastructurally. *Brightman* et al. [3] demonstrated opening of tight junctions to peroxidase tracer after intracarotid infusion of 3 *M* urea. When endothelial cells shrank because of the hypertonic environment, the membranes stressed the tight junctions and made them permeable to intravascular tracer. Separation of cerebral endothelial tight junctions of 0.1 μm was observed in rats after intracarotid infusion of arabinose [9].

Reversibility of barrier disruption has been documented by the administration of Evans blue-albumin at different times after intracarotid hypertonic solution infusion. After disruption, the Evans blue staining was not evident when this dye was given 30 min after mannitol infusion [12, 29]. Thus, for this 68,000-molecular-weight marker, the barrier remains open for less than 30 min. However, slight permeability changes have been shown for other markers for up to 2 h after barrier disruption [12, 13, 30]. The metabolic effects of hypertonic disruption of the BBB include transient increase of glucose metabolism and transient increase of brain water content by 1.5% [25]. The slight increase in brain water (cerebral edema) requires up to 24 h to resolve [30].

Several hypertonic solutions have been used to disrupt the BBB, including arabinose, mannitol, and radiographic contrast agents. They are all effective, but have minor disadvantages. Mannitol (25%) is a nearly saturated solution at ambient temperature and must be filtered before use. Direct installation of certain iodinated radiographic contrast agents into the internal carotid artery causes EEG changes that may persist for more than 2 h [27].

Pappius et al. [25] observed conscious rats during and after osmotic disruption of the BBB. Transient behavioral changes, such as turning to one side and depressed activity, were seen. *Rapoport* et al. [28] did not observe neurologic or behavioral sequelae in 7 of 8 rhesus monkeys given intracarotid infusions of 2.5 *M D,L*-lactamide; the eighth animal developed a right hemiparesis. After 48 h, the monkeys were sacrificed. No cerebral edema was evident nor were there significant changes in brain water, Na^+, or K^+ content. However, intraocular pressure was reduced and there was histologic evidence of severe damage to the ciliary epithelium of the eye [24]. 18 months later, at least some visual acuity was retained [*Rapoport,* personal commun.]. In the rat, no such pathologic eye changes have been observed [*Rapoport,* personal commun.].

With reversible osmotic barrier disruption, permeability increases about 20-fold for substances that the brain normally excludes, such as sucrose, microperoxidase, penicillin, gamma globulin, and neurotransmitters [27, 29].

BBB Disruption in Dogs

Recently reported studies of reversible osmotic BBB disruption in the dog [12, 13] provide a working model that has significant advantages over the previous small-animal murine model [27, 29]. The dog model permits an opportunity to obtain serial samples of CSF in sufficient quantity; secondly, the brain of dogs, in contrast to that of the rat, is large enough to be evaluated by computerized tomography (CT), a monitor of BBB disruption after intravenous administration of iodinated contrast material and a noninvasive criterion of the success of the disruption. Thirdly, dogs are readily available for experimental purposes and are less costly than primates.

The dog, unlike the rat, has a small internal carotid artery (2 mm) located distally just under the mandible, and a large distal external-to-internal carotid collateral circulation. In the early studies in dogs, hypertonic (25%) mannitol was infused through the common carotid artery, but

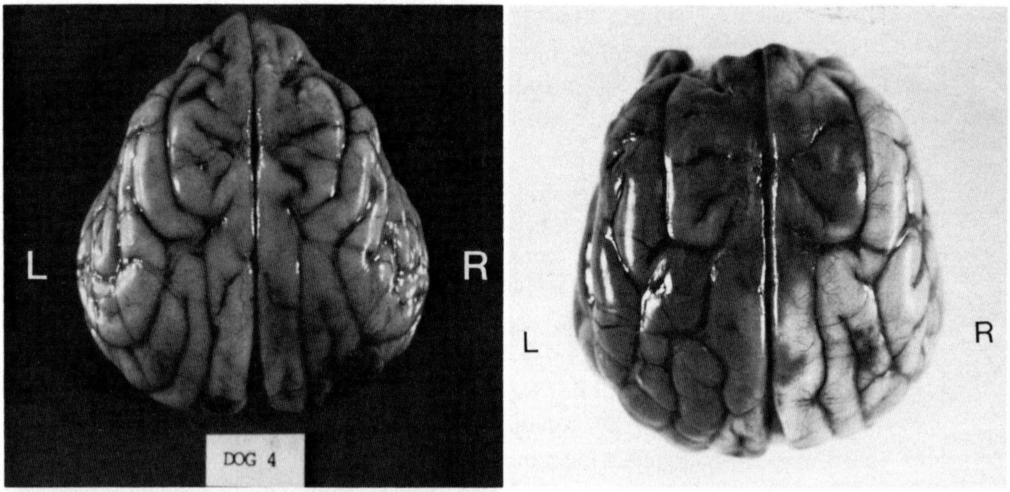

a b

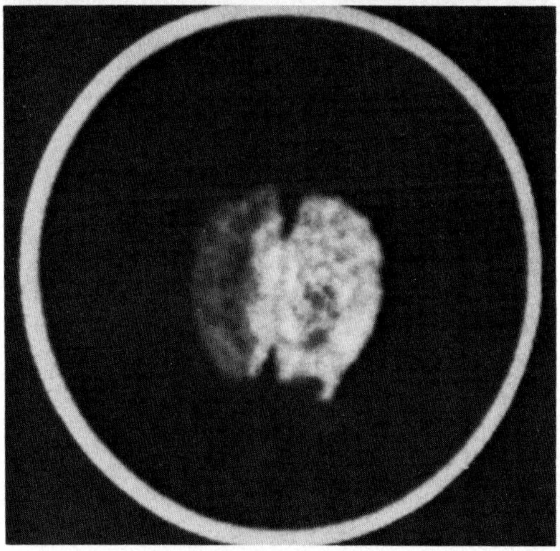

c

Fig. 1. *a* Unfixed canine brain 1 h after the infusion of saline rather than mannitol into the left internal carotid artery. There is no staining of the cerebral parenchyma, although Evans blue-albumin was given before the infusion of saline. *b* Unfixed canine brain 1 h after osmotic BBB disruption with 25% mannitol via the left internal carotid artery. Evans blue dye (molecular weight 68,500) was given intravenously before the intraarterial mannitol infusion. The dark areas represent penetration of Evans blue-albumin across the BBB. The BBB was disrupted in the entire left cerebral hemisphere and in the distribution of the anterior cerebral artery in the right hemisphere. *c* CT scan of canine brain after osmotic disruption of the BBB with intracarotid mannitol. Barrier disruption was preceded by an injection of an iodinated radiographic contrast agent (Conray 60, 3 ml/kg) and of Evans blue dye. The entire left cerebral hemisphere is markedly enhanced (appears white), as well as the brain in the right hemisphere which is in the distribution of the right anterior cerebral artery. This distribution of enhancement was an exact reflection of BBB disruption as indicated by transudation of Evans blue-albumin. The disruption of the brain in the distribution of the right anterior cerebral artery is almost certainly the result of a patent anterior communicating artery.

this resulted in variable BBB disruption, as determined by transudation of Evans blue given intravenously [14] and by variable levels of MTX in the brain. It is now clear that, in this model, mannitol infused into the common carotid artery produces inconsistent blanching of the ipsilateral cerebral cortex [14].

A major technical advance was made by simply altering the site of infusion. When the distally located internal carotid artery was cannulated, 42 ml of fluid infused over 30 s completely and consistently blanched the cerebral cortex of 25-kg mongrel dogs. When infusion was into the internal carotid artery rather than the common carotid artery, the levels of MTX in brain rose from 1,000–2,000 ng of MTX/g of tissue to as high as 90,000 ng of MTX/g of tissue. MTX levels in brain tissue of control animals receiving internal carotid artery infusion of saline were about 1,000 ng/g. The staining of the brain parenchyma following intravenous Evans blue infusion was also much more pronounced and consistent than it had been with common carotid artery infusion (fig. 1). MTX levels in the CSF more accurately reflected the levels in brain with the internal carotid infusion method than with the common carotid artery technique; nevertheless, the correlation is not adequate to use this measurement as a parameter of pharmacologic planning.

It should be noted that *Ohno* et al. [23] have observed a 10- to 50-fold increase of MTX levels in brain in the rat model after osmotic BBB disruption. Similar results have been reported by *Hasegawa* et al. [7].

Studies of the Use of CT as a Means of Monitoring BBB Disruption in the Dog

CT has been well established as a sensitive method for detecting minute changes in the radiographic density of soft tissues. Differences in X-ray attenuation of 1% or less can be seen with the CT scan method.

Brain lesions that disrupt the BBB, such as CNS tumors, infarcts, or infections, can be more easily seen and more accurately located after the intravenous injection of iodinated contrast material. Normally, these iodinated, water-soluble agents are not able to penetrate the BBB [12]. After disruption of the BBB surrounding CNS lesions, however, small amounts of the contrast agent extravasate from the intravascular space into the interstitium of the affected area. Because the CT scanner is approximately 100 times more sensitive to the detection of iodine molecules than are conventional X-ray techniques, the result is a significant increase in attenuation of X-ray transmission in these areas on the CT scan image; this is referred to as contrast enhancement within a lesion. A logical extension of these observations is that the CT scan could be applied to provide a radiographic image to document the presence of BBB disruption as well as to afford a means of mapping the precise areas of the brain that have been affected. Furthermore, in addition to producing images of superior quality, this method produces numerical values, the 'CT number', that correlate with the degree of barrier disruption and should therefore provide a parameter by which to quantitate access of agents (MTX, adriamycin, and alkylating agents) across the BBB.

In studies of BBB disruption in the dog model [12], CT scan images revealed contrast enhancement of the brain parenchyma in the regions where the barrier had been disrupted (fig. 1). The areas of the brain that were stained with Evans blue dye showed excellent correlation with the enhanced regions on the CT scan. The attenuation differences between the normal areas of brain and the areas that had been disrupted and enhanced with iodine showed a difference between the CT numbers of 20 units, which corresponds to a 2% difference in attenuation relative to that of water content and is easily detected on the CT scanner. There is a linear relation between change in CT number and iodine content in tissue.

Osmotic BBB Disruption in the Posterior Fossa of the Dog

Transient reversible osmotic BBB disruption was produced in the posterior fossa of 33 dogs [16] using a new percutaneous catheter technique, rather than the open surgical technique, that was developed for the infusion of the hypertonic mannitol into the vertebral artery. This method now makes it possible to obtain serial barrier disruption in the same animal. Neither the catheter technique nor the osmotic barrier modification resulted in interference with brain stem function in most of these dogs. The degree of barrier modification achieved by osmotic BBB disruption in the posterior fossa was proportional to that described earlier for barrier modification of the supratentorial parenchyma. MTX (100 mg) delivered to the brain via the vertebral artery resulted in a drug concentration in the range of 100–300 ng MTX/g of brain tissue. When the same amount of drug was given following osmotic BBB disruption, brain tissue concentration of MTX was 2,000–3,000 ng/g of brain tissue. In addition, BBB modification was quantitated and monitored by the amount of enhancement observed on CT scans.

Effects of Adrenal Corticosteroids on Drug Delivery to the Brain

The effect of adrenal cortical steroids on cytoreductive drug delivery to a glioma model in rats is being studied [17]. The results in rats with avian sarcoma virus-induced glioma showed that the administration of adrenal steroids in doses equivalent to those used in humans with brain tumors without BBB modification resulted in mildly reduced delivery of MTX to both the tumor and the brain adjacent to tumor. Following osmotic BBB disruption, MTX delivery to the tumor-bearing hemisphere was increased; however, when adrenal steroids were given before the barrier modification and MTX therapy, the level of MTX was significantly less in the tumor and brain around the tumor. These studies provide evidence that the BBB exists in tumors, that it is a factor in the delivery of drug to tumors, and that steroid administration can significantly decrease the amount of MTX delivered to the tumor and brain around tumor.

Pharmacology and Toxicity of Intracarotid Adriamycin Administration after Osmotic BBB Modification

The effect of reversible BBB modification on the delivery of adriamycin to the brain was studied in the rodent and canine models [18]. Pharmacokinetic and toxicologic studies were done in these animals after they received a wide range of doses of adriamycin (0.1–1.0 mg/kg), administered into the carotid artery following osmotic BBB modification with mannitol. When BBB modification was not induced, no immunoreactive adriamycin was detected in the cerebrum; whereas following barrier modification, up to 4.5 μg of drug and/or metabolites per gram of brain tissue were found by radioimmunoassay technique. Optimum levels of adriamycin and metabolites were achieved in the tissue after BBB modification when adriamycin was administered either by rapid bolus injection over 30 seconds or by slow, continuous infusion over 15 min. Immunoreactive adriamycin was identified in brain for up to 6 h after administration. Significant functional neurotoxicity occurred at all dosage levels – even at 0.1 mg/kg, a level at which adriamycin concentration in the brain was below the level of detectability. Neuropathologic examination revealed the presence of necrosis and hemorrhagic infarcts in animals receiving adriamycin after BBB disruption. In control animals in which the barrier was disrupted but no adriamycin was given and in those given adriamycin but saline instead of mannitol (i.e., those with no BBB disruption), no

CNS toxicity or damage was observed. These pharmacologic and toxicity studies suggest that adriamycin or its metabolites may produce significant clinical neurotoxicity in the canine model when even small amounts penetrate the BBB.

Studies are currently under way to review the neuropathologic effects and pharmacokinetics of bleomycin, MTX, cyclophosphamide, cisplatinum, mitomycin C and 5-fluorouracil (5-FU) after BBB modification. These studies are only beginning, but preliminary results suggest that Cytoxan and MTX are not toxic, that bleomycin and mitomycin C have some toxicity, and that cis-platinum and 5-FU are very toxic.

Inability of Dimethyl Sulfoxide and 5-FU to Modify the BBB

It has been suggested that the intravenous administration of dimethyl sulfoxide (DMSO) or 5-FU can accomplish the same reversible barrier disruption as does the osmotic technique, but in a less invasive manner. More specifically, *Broadwell and Salcman* [4] have suggested that intravenously administered DMSO can disrupt the BBB and increase permeability of chemotherapeutic agents across the BBB. *MacDonell* et al. [11] have suggested that intravenous 5-FU also can modify the BBB. We have been unable to modify the BBB significantly with either of these drugs [19]. DMSO was administered intravenously to 25 rats at concentrations ranging from 25 to 90%, and the penetration of MTX, Evans blue-albumin, and hexosaminidase-A was evaluated at times ranging from 1.5 to 3.5 h. No significant BBB modification was observed. In a second series of animals, intravenous (2.5 ml of 15% or 30% DMSO) and intraperitoneal (5 ml of 15% or 30% DMSO) DMSO was administered. Again no modification of the blood brain barrier was observed, although 2 animals in this group developed seizures. Finally, DMSO was also given intraarterially via the external carotid artery of the rat into the internal carotid artery. BBB modification, albeit generally modest, was obtained in the rats receiving intracarotid DMSO, but this may have been attributable to grand mal seizures, as 5 of 6 of these animals had such seizures. Several of the rats receiving DMSO intravenously had seizures; all of these animals developed varying degrees of hematuria. *Grieg and Rapoport* (personal communication) have determined the permeability × capillary surface area (PA) values for horseradish peroxidase and melphalan in rats after a combination of intravenous and intraperitoneal DMSO administration. There was no change in PA values in animals receiving DMSO when compared with control animals. Similarly, 5-FU was administered intravenously at a dosage of 30 mg/kg, and the permeability of the BBB to either Evans blue-albumin or MTX was evaluated after 6.5 h, as suggested by *MacDonell* et al. [11]. No increased permeability of the BBB to these two markers was observed. In summary, osmotic BBB disruption, in our experience, is still the most consistent and reliable means available to modify the BBB in a reversible fashion. Neither intravenously administered DMSO nor 5-FU appears to improve chemotherapy or protein tracer delivery to the CNS, and the use of DMSO can result in seizures and hematuria.

Delivery of Protein across the BBB

Studies have also been undertaken to evaluate the possibility that hexosaminidase A, the enzyme deficient in Tay-Sachs disease, could be delivered effectively to brain [1]. By using the rat model of BBB modification and radiolabeled enzyme, increased delivery of hexosaminidase A to brain has been demonstrated in more than 85 rats [20]. The time of injection of hexosaminidase A after BBB disruption is critical for achieving maximum delivery. Rats receiving a rapid, 30-second bolus intraarterial administration of hexosaminidase A immediately after BBB disruption showed a marked increase in enzyme delivery to the brain when compared with controls that did not have prior barrier disruption. When the enzyme was administered 15–20 min after BBB disruption, 50% less hexosaminidase A was delivered; when the enzyme was given 60–120 min after BBB modification, the amount delivered was the same as in control animals. This critical time course is very different from that seen in trials of the low molecular weight chemotherapeutic agents MTX and adriamycin.

Our more recent studies focus on the capacity of hexosaminidase A to enter brain cells after BBB disruption and the fate of the enzyme upon its entry into the cell. To conduct these studies, an antibody to human hexosaminidase A that does not cross-react with the rodent enzyme was prepared. Using this antibody, it was possible to document that human hexosaminidase A administered via the internal carotid artery after BBB modification traversed the barrier, and once across the barrier, was taken up by cells. Latency studies indicate that up to 75% of the enzyme that was delivered across the BBB (0.2–0.5% of the administered dose) was released simultaneously with the lysosomal marker beta galactosidase.

These observations demonstrate that human hexosaminidase A is deposited in subcellular organelles, possibly the lysosome, after delivery across the BBB and remains enzymatically active. Thus, enzyme replacement may be a feasible therapeutic approach to the treatment of Tay-Sachs disease. With regard to brain tumors, we are using the knowledge we obtained with hexosaminidase A to deliver tumor-specific monoclonal antibodies across the BBB.

Clinical Studies of Osmotic BBB Modification in Patients with Malignant Brain Tumor

In our initial clinical studies, MTX was administered 33 times to 6 patients with brain tumors after BBB disruption. No permanent complications were observed [15]. Serial enhanced CT scans and quantification by CT number indicated that disruption increased drug delivery to the tumor and the immediately surrounding brain (fig. 2). Neuroradiologic evaluation showed that drug persisted in the tumor longer after barrier disruption than when it was delivered without prior BBB disruption. The concentration of MTX in spinal fluid did not correlate with the degree of barrier disruption measured by CT and radionuclide scans. In 1 patient, an anatomic variation in the circle of Willis resulted in barrier disruption extending into the posterior fossa without ill effect. Osmotic BBB disruption appears to be a nontoxic procedure in humans, and is a means to increase drug delivery to both malignant brain tumors and surrounding brain parenchyma.

In our more recent clinical trials, we have documented tumor regression in patients who have microglioma, medulloblastoma, and glioblastoma [21]. However, with increasing experience and success in opening the BBB using intracarotid mannitol, there has been a concomitant increased incidence of focal motor and grand mal seizures. These seizures appear to be a direct result of the use of meglumine iothalamate, a known epileptogenic iodinated contrast agent. This has been the main complication of osmotic BBB modification followed by chemotherapy. As a result, we recently have been relying on radionuclide scanning, an adequate and safer, though less sensitive, means to monitor barrier modification. By employing this method, it appears that we have mostly eliminated the incidence of seizures after osmotic BBB modification and chemotherapy administration.

Before concluding this review, it is worth mentioning that 2 patients

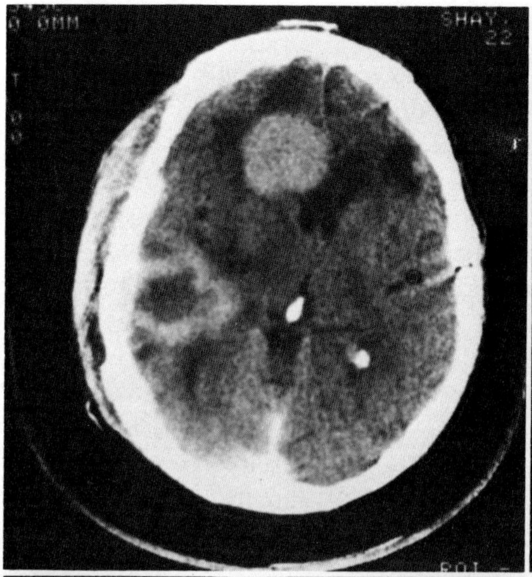

a

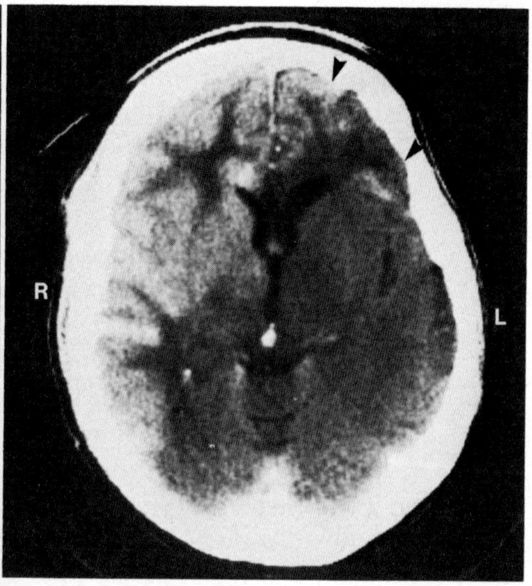

b

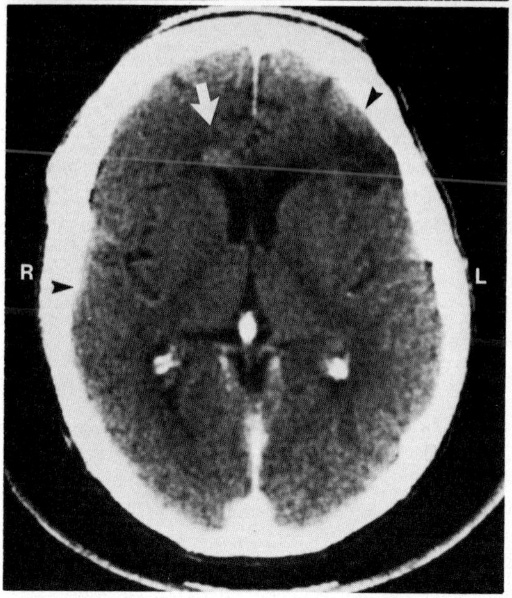

c

Fig. 2. Serial enhanced CT scans in a patient with a diagnosis of primary CNS lymphoma. *a* CT scan after the patient's second craniotomy while being treated with Decadron (24 mg/day) just before entering the experimental protocol. Three lesions are observable on this transverse cut CT scan at the level of the lateral ventricles: a right frontal, uniformly enhancing mass; a ring lesion in the right posterior temporal region at the site of the second craniotomy, and a low-density lesion in the left frontal region which was the site of the initial tumor excision. (Black square in left cerebral hemisphere is a CT marker and should be disregarded.) *b* Transverse cut enhanced CT scan at the level of the lateral ventricles 30 min after the third BBB modification. The white areas depict the areas of iodinated contrast penetrating the BBB in the distribution of the anterior and middle cerebral arteries. It should be noted that BBB modification also occurred in the anterior cerebral distribution on the left side (arrowheads) owing to collateral flow via the anterior communicating artery. There is barrier modification in and around all three known tumor regions. The CT number in the right frontal tumor increased from 54 on a control enhanced CT scan 24 h before BBB opening to 75 after BBB opening; the time between intravenous contrast administration and scanning and the contrast dose was the same for both scans. Both this scan and the control scan (not shown) were obtained 5 days after discontinuation of steroid treatment. *c* CT scan at the level of the lateral ventricles just before the fifth course of therapy. Disappearance of the right posterior temporal lesion (arrowhead) and marked diminution in the size of the right frontal lesion (white arrow) is seen. The low-density area that does not enhance at the site of the previous surgery in the left frontal region persists and is probably a postoperative cavity (arrowhead) (fig. *a* and *b* are reproduced with permission [22]).

with primary CNS lymphoma (fig. 2) have had complete remission and a near complete remission, respectively, with multiagent chemotherapy administered in association with reversible BBB modification used to enhance drug delivery to the tumor [22]. Radiographic CT studies in association with serial barrier disruption document the propensity of steroids to alter the radiographic pattern of the tumor in hours. In addition, monitoring of BBB disruption with the aid of enhanced CT scans provided visualization of tumor and its extent that was not apparent by conventional means, and the scans further supported the existence of a BBB in CNS tumors. These studies also provide evidence that drug delivery to the tumor, as well as to surrounding brain, is enhanced following reversible BBB modification, and that such therapy results in a defined objective response. Finally, for the first time, BBB modification in the posterior fossa was induced without untoward effects by the infusion of mannitol into the vertebral artery.

Acknowledgments

These studies were supported by the Veterans' Administration, the John Raaf Institute, PHS Grant CA31770, and the Oregon Medical Research Foundation, as well as NIH Grant CA23115, the Meadows Foundation, and the Southwestern Medical Foundation-Kinsler Williamson Brown Fund.

The author would like to thank Dr. *Stanley Rapoport* and Dr. *Gene Frenkel* for their help and advice. The technical assistance of Ms. *Peggy Barnett,* Ms. *Suellen Hill,* RN, Mr. *Mike Pagel,* and Mr. *Chris McCormick* was greatly appreciated.

References

1 Barranger, J.A.; Rapoport, S.I.; Fredericks, W.; MacDermot, K.: Modification of the blood brain barrier: increased concentration and fate of enzymes entering the brain. Proc. natn. Acad. Sci. USA *76:* 481–485 (1978).

2 Benjamin, R.S.; Wiernik, P.H.; Bachar, N.R.: Adriamycin chemotherapy – efficacy, safety, and pharmacologic basis of intermittent single high dose schedule. Cancer *33:* 19–27 (1974).

3 Brightman, M.W.; Hori, M.; Rapoport, S.I.; Reese, T.S.; Westergaard, R.: Osmotic opening of tight junctions in cerebral endothelium. Com. Neur. *152:* 317–326 (1973).

4 Broadwell, R.D.; Salcman, M.: Expanding the definition of the blood-brain barrier to protein. Proc. natn. Acad. Sci. USA, Neurobiol. *12:* 7820–7824 (1981).

5 Fenstermacher, J.D.; Johnson, J.A.: Filtration and reflection coefficients of the rabbit blood-brain barrier. J. Physiol., Lond. *211:* 341–346 (1966).

6 Groothuis, D.R.; Fischer, J.M.; Lapin, G.; Bigner, D.D.; Vick, N.A.: Permeability of different experimental brain tumor models to horseradish peroxidase. J. Neuropath. exp. Neurol. *41:* 164–185 (1982).

7 Hasegawa, H.; Allen, J.C.; Mehta, B.M.; Shapiro, W.R.; Posner, J.B.: Enhancement of CNS penetration of methotrexate by hyperosmolar intracarotid mannitol or carcinomatous meningitis. Neurology *29:* 1280–1286 (1979).

8 Kimelberg, H.K.; King, D.; Watson, R.E.; Reiss, F.L.; Biddlecome, S.M.; Bourke, R.S.: Direct administration of methotrexate into the central nervous system of primates. Part 1: Distribution and degradation of methotrexate in nervous and systemic tissue after intraventricular injection. J. Neurosurg. *48:* 883–894 (1978).

9 Levin, V.A.; Clancy, T.P.; Ausman, J.I.; Rall, D.P.: Uptake and distribution of ^{3}H-methotrexate by the murine ependymoblastoma. Natn. Cancer Inst. Monogr. *48:* 875–883 (1972).

10 Long, D.M.: Capillary ultrastructure in human metastatic brain tumors. J. Neurosurg. *51:* 53–58 (1979).

11 MacDonell, L.A.; Potter, P.E.; Leslie, R.A.: Localized changes in blood-brain barrier permeability following the administration of anti-neoplastic drugs. Cancer Res. *38:* 2930–2934 (1978).

12 Neuwelt, E.A.; Maravilla, K.R.; Frenkel, E.; Barnett, P.; Hill, S.; Moore, R.: The use of enhanced computerized tomography to evaluate osmotic blood brain barrier disruption. Neurosurgery *5:* 576–582 (1979).

13 Neuwelt, E.A.; Maravilla, K.R.; Frenkel, E.P.; Hill, S.; Rapoport, S.I.; Barnett, P.: Osmotic blood-brain barrier disruption: computerized tomographic monitoring of chemotherapeutic agent delivery. J. clin. Invest. *64:* 684–688 (1979).

14 Neuwelt, E.A.; Frenkel, E.P.; Diehl, J.; Maravilla, K.; Vu, L.; Clark, K.; Rapoport, S.; Barnett, P.; Hill, S.; Lewis, S.E.; Ehle, A.L.; Beyer, C.W., Jr.; Moore, R.J.: Osmotic blood brain barrier disruption: a new means of increasing chemotherapeutic agent delivery. Trans. Am. neurol. Ass. *104:* 1–5 (1979).

15 Neuwelt, E.A.; Frenkel, E.P.: Monitoring methotrexate delivery in patients with malignant brain tumors after osmotic blood brain barrier disruption. Ann. intern. Med. *94:* 449–454 (1981).

16 Neuwelt, E.A.; Diehl, J.; Frenkel, E.P.; Barnett, P.: Osmotic blood brain barrier disruption in posterior fossa of the dog. J. Neurosurg. *55:* 742–748 (1981).

17 Neuwelt, E.A.; Bigner, D.; Frenkel, E.P.; Barnett, P.; Pagel, M.: The effects of osmotic modification of the blood brain barrier and adrenal steroid administration on methotrexate delivery of gliomas in rats: the blood brain barrier is a factor. Proc. natn. Acad. Sci. USA. *79:* 4420–4423 (1982).

18 Neuwelt, E.A.; Glasberg, M.; Frenkel, E.P.; Pagel, M.; Barnett, P.: Pharmacology and toxicity of intra-carotid adriamycin administration following osmotic barrier modification. Cancer Res. *41:* 4466–4470 (1981).

19 Neuwelt, E.A.; Barnett, P.; Barranger, J.; McCormick, C.; Pagel, M.; Frenkel, E.: The inability of dimethyl sulfoxide (DMSO) and 5-fluorouracil (5-FU) to open the blood brain barrier. Neurosurgery *12:* 29–34 (1983).

20 Neuwelt, E.A.; Barranger, J.A.; Brady, R.O.; Pagel, M.; Furbish, F.S.; Quirk, J.M.; Mook, G.E.; Frenkel, E.: Delivery of hexosaminidase A to the cerebrum after osmotic modification of the blood-brain barrier. Proc. natn. Acad. Sci USA. *78:* 5838–5841 (1981).

21 Neuwelt, E. A.; Howieson, J.; Specht, D.; Haynes, J.; Bennett, M.; Hill, S.: Comparison of enhanced computerized tomography and radionuclide brain scan to monitor osmotic blood brain barrier disruption. Am. J. Neuroradiol. *4:* 907–913 (1983).

22 Neuwelt, E. A.; Frenkel, E.; Diehl, J.; Hill, S.; Balaban, E.: Successful treatment of primary central nervous system lymphomas (microgliomas) with chemotherapy after osmotic blood brain barrier opening. Neurosurgery *12:* 662–671 (1983).

23 Ohno, K.; Fredericks, W. R.; Rapoport, S. I.: Osmotic opening of the blood-brain barrier to methotrexate in the rat. Surg. Neurol. *12:* 323–328 (1979).

24 Okisaka, S.; Kuwabara, T.; Rapoport, S. I.: Selective destruction of the pigmented epithelium in the ciliary body of the eye. Science *184:* 1298–1299 (1974).

25 Pappius, H. M.; Savake, H. E.; Feischi, C.; Rapoport, S. I.; Sokoloff, L.: Osmotic opening of blood-brain barrier and local cerebral glucose utilization. Ann. Neurol. *5:* 211–219 (1979).

26 Posner, J. B.: Management of central nervous system metastases. Semin. Oncol. *4:* 81–91 (1977).

27 Rapoport, S. I.: Blood-brain barrier in physiology and medicine (Raven Press, New York 1976).

28 Rapoport, S. I.; Matthews, K.; Thompson, H. K.; Pettigrew, K. D.: Osmotic opening of the blood-brain barrier in the rhesus monkey without measurable brain edema. Brain Res. *136:* 23–29 (1977).

29 Rapoport, S. I.; Ohno, K.; Fredericks, W. R.; Pettigrew, K. D.: Regional cerebrovascular permeability to ^{14}sucrose after osmotic opening of the blood-brain barrier. Brain Res. *150:* 653–657 (1978).

30 Rapoport, S. I.; Fredericks, W. R.; Ohno, K.; Pettigrew, K. D.: Quantitative aspects of reversible osmotic opening of the blood-brain barrier. Am. J. Physiol. *238:* R 421–431 (1980).

31 Rosen, G.; Ghavimi, F.; Nirenberg, A.; Mosende, C.; Mehta, B. M.: High-dose methotrexate with citrovorum factor rescue for the treatment of central nervous system tumors in children. Cancer Treatment Rep. *61:* 681–690 (1977).

32 Shapiro, W. R.; Mehta, B.; Blasberg, R. G.; Patlak, C. S.; Kobayashi, T.; Allen, J. C.: Pharmaco-dynamics of entry of methotrexate into brain of humans, monkeys and a rat brain tumor model; in Paoletti et al. (eds.) Proc. Int. Symp. on Multidisciplinary Aspects of Brain Tumor Therapy (North-Holland, Elsevier, Amsterdam 1979).

33 Tator, C. H.: Chemotherapy of brain tumors: uptake of tritiated methotrexate by a transplantable intracerebral ependymoblastoma in mice. J. Neurosurg. *37:* 1–8 (1972).

34 Vick, N. A.; Bigner, D. D.: Chemotherapy of brain tumors: the blood-brain barrier is not a factor. Archs Neurol. *34:* 523–526 (1977).

35 Walker, M. D.; Weiss, H.: Chemotherapy in the treatment of malignant brain tumors. Adv. Neurol. *13:* 149–191 (1975).

E. A. Neuwelt, MD, Departments of Surgery (Division of Neurosurgery) and Biochemistry, Oregon Health Sciences University, Portland, OR 97201 (USA)

Prog. exp. Tumor Res., vol. 28, pp. 67–78 (Karger, Basel 1984)

Megadose Chemotherapy with Bone Marrow Rescue

H. J. Fingert, F. H. Hochberg

Hematology-Oncology Unit and Neurology Service, Massachusetts General Hospital and Sidney Farber Cancer Institute, Harvard Medical School, Boston, Mass., USA

Rationale for Megadose Chemotherapy

Although nitrosoureas have had great promise for brain tumor therapy in animal studies because of their ability to penetrate the blood-brain barrier, the clinical usefulness of these drugs at standard doses in the management of primary or metastatic brain tumors has been limited. There are reports of patients with malignant gliomas who have long-term survival after multimodal therapy [6], but these are rare. The median time to tumor progression for most patients with recurrent malignant gliomas is 6 months [23].

The limitation of nitrosourea utility is characteristic, delayed, and cumulative dose-related myelosuppression that limits both the total dosage and number of treatments patients receive. As a result, fewer than 40% of patients receive as much as 1,200 mg 1,3-bis(2-chloroethyl)-1-nitrosourea (BCNU)/m^2 total dose during their entire treatment period (four to five courses at 240 mg/m^2 per course) [36]. Delayed myelosuppression limits dose frequency to 6- to 8-week periods.

The therapeutic advantage of achieving maximum doses of chemotherapy has been demonstrated in some solid tumors and hematologic malignancies [16]. Few controlled clinical trials have been reported that compare the effects of nitrosoureas at different dosage levels in cancer therapy, but a dose-response relationship has been suggested by retrospective data from cases of primary brain tumors [24] and systemic malignancies [7]. For nitrosoureas, a steep dose-response effect has been demonstrated in experimental animals with implanted intraperitoneal and intracerebral tumors [31]. In rat 9L gliosarcomas, single BCNU doses of

13.3 mg/kg 600 mg/m^2 produce a survival rate twice that obtained when the same total dose is divided over 2 days [28]. A steep dose-response curve for BCNU has been demonstrated also in tissue culture cell lines derived from human brain tumors, although these heterogeneous lines may have cells of varying drug sensitivities [22, 29, 38]. These data suggest that if the dose-limiting myelosuppression of nitrosoureas could be overcome, then larger than usual doses – 'megadoses' – of drug could improve the response rate and survival in patients with primary brain tumors.

Rationale for Bone Marrow Rescue

Several centers are treating patients who have refractory malignancies by means of megadose chemotherapy followed by intravenous autologous bone marrow transplantation (ABMT) to prevent hematopoietic toxicity.

Animal studies in rodents, dogs, and monkeys indicate that infusions of stored autologous marrow cells are effective in preventing death from myelosuppression following treatment with a variety of alkylating agents and irradiation [8]. In dogs, protection against the lethality of high-dose BCNU from myelosuppression has been achieved with ABMT [1, 26], although nonmarrow BCNU toxicity produced death from gastrointestinal and pulmonary causes. In addition to reduced lethality from myelosuppression, animal trials have demonstrated that tumor response and survival rates were better following high-dose chemotherapy or total-body irradiation with ABMT than after therapy with lower doses given without marrow rescue [12, 15, 37].

Early clinical trials using ABMT rescue after megadose chemotherapy demonstrated the safety of autologous marrow transplantation and the feasibility of using the large drug dosages, but were inconclusive with regard to the benefit of ABMT because there had been no control groups receiving megadose chemotherapy alone [8]. Recent clinical studies have shown that the kinetics of marrow recovery in patients given autologous bone marrow rescue therapy after high-dose chemotherapy was better than that in control patients who did not have ABMT [2, 17].

Several protocols for solid tumors have included megadose BCNU (600–2,850 mg/m^2) as a single agent or standard-dose BCNU in a combination regimen followed by bone marrow rescue. Although some centers report a high response rate, their figures often reflect a greater than true

Table I. Megadose chemotherapy with bone marrow rescue in solid tumors

Tumor	Regimen	Responses[1]/ evaluable cases	Duration months	Reference
Lymphoma	multidrugs + BCNU	13/13	1–40+	[3, 17]
Melanoma	BCNU	14/30	2–4+	[13, 14]
Melanoma	melphalan, BCNU	8/13	1–18+	[35]
Melanoma, oat cell	BCNU	1/9	6	[33]
Testicular	multidrugs + BCNU	7/10	1–5+	[4]
Mixed	multidrugs + BCNU	18/29	1–16+	[9, 33]
Mixed	multidrugs + BCNU	9/11	1–29+	[10, 30]
Mixed	multidrugs	2/7	16–25+	[21]
Mixed	multidrugs	12/13	1–16+	[11]

[1]Complete and partial remissions combined.
Criteria for responses appear in each publication.

response rate among all treated patients because early deaths frequently are excluded in these calculations. The most notable systemic responses and/or long-term remissions reported have been obtained using megadose chemotherapy-ABMT as primary or secondary therapy for Burkitt's lymphoma, melanoma, testicular carcinoma, neuroblastoma, and ovarian carcinoma. Several of these studies have used ABMT with BCNU alone or BCNU in combination with other drugs, although others used ABMT following other megadose drugs, including melphalan, adriamycin, and cyclophosphamide. Studies differ with respect to the use of refrigerated or frozen ABMT, as well as in the intervals between marrow harvest and infusion. Although they suggest that marrow remains viable in storage postharvest from 24 h to 4 days if kept refrigerated or from 24 h to years if kept frozen, there is no definite evidence to point to greater efficacy for one or another preparative/storage technique.

Table I summarizes several recently reported trials in patients with solid tumor metastases to the central nervous system (CNS) that were found either at the time of initial therapy or as a site of relapse after a prior systemic response to megadose chemotherapy and ABMT.

Appelbaum et al. [3] reported 1 patient with cerebrospinal fluid cytology results diagnostic of Burkitt's lymphoma who achieved a long-term (29+ months) complete remission after megadose multidrug chemotherapy that included BCNU, 200 mg/m^2, and ABMT rescue. *Ekert* et al. [11] reported a pediatric patient who had CNS and bone marrow metastases 6 months after apparently complete remission that had been induced by combination systemic chemotherapy-ABMT and prophylactic intrathecal methotrexate. Another patient with retinoblastoma attained a complete systemic remission after combination chemotherapy-ABMT, but had metastasis to the CNS after 5 months. CNS responses have been noted after BCNU-ABMT in patients with oat cell carcinoma also [33].

Fay and Phillips [14] reported 18 patients with CNS metastases from a variety of solid tumors who received BCNU, 1,200–2,850 mg/m^2, and ABMT. Half of these patients were responders, as documented by CT scan and/or cerebrospinal fluid examination, including 1 who had complete remission and 3 patients with metastatic melanoma who had partial remissions. However, brain metastases from melanoma remain unresponsive in other series: *Thomas* et al. [35] treated 13 patients with systemic melanoma using ABMT rescue with nitrogen mustard, melphalan, or BCNU (400–800 mg/m^2). Despite 2 patients who had a brief partial remission (6 and 10 weeks, respectively), the most frequent cause of death in this series was intracerebral metastases (in 7 of 8 patients).

Based on these preliminary results in refractory solid tumors, a collaborative study has been undertaken at the Massachusetts General Hospital and Sidney Farber Cancer Institute in Boston using megadose BCNU with ABMT in patients with recurrent glioblastoma. Glioblastoma patients fulfill most of the criteria suggested as indicating a likely favorable response to treatment with megadose chemotherapy and ABMT [8]: (1) a strong dose-response effect of nitrosoureas is predicted by prior clinical, animal, and in vitro tissue culture studies; (2) myelosuppression is the major dose-limiting nitrosourea toxicity; (3) the relatively young age (median 50 years) and good performance status (Karnofsky scale, 80%) of patients eligible for our study imply that marrow harvest and infusion can be accomplished safely; and (4) it is extremely unusual for glioblastomas to metastasize to the bone marrow.

Methods: Megadose BCNU and Bone Marrow Rescue for Recurrent Glioblastoma

18 patients with postoperative and postirradiation recurrent glioblastoma were treated with megadose BCNU given in progressively higher

doses and ABMT. Patients ranged from 17 to 57 years of age. 6 patients had received adjuvant therapy with N-(2-chloroethyl)-Nα-cyclohexyl-N-nitrosourea (CCNU) before entering into this study. Indicators of response included the improvement or worsening of the patient's Karnofsky score and the results of contrast-enhanced CT while the patient was receiving isodose or diminished doses of corticosteroids. Weekly blood counts and liver function studies and periodic chest X-rays and tests of pulmonary and renal function were performed also.

With the patient under general anesthesia, bone marrow was obtained by 2 physicians over a period of 1.5 h [20]. The marrow was mixed with adenine-citrate-dextrose solution, filtered and centrifuged to remove fat and bone spicules, and stored at 4°C for 48–72 h until infusion. Marrow stem-cell viability, measured by in vitro colony-forming ability, was over 90% during this storage period.

BCNU was administered in divided doses on the day of marrow procurement (4 h after harvest) and again 1 day later (approximately 20 h after harvest). Each patient received a total dose of BCNU ranging from 600 to 1,400 mg/m^2. 6 patients who had received prior and adjuvant CCNU were given 800 mg BCNU/m^2. 5 patients receiving 1,200 mg/m^2 were given prophylactic oral antibiotics (nystatin and trimethoprim-sulfamethoxazole) during the period of severe neutropenia. 3 patients subsequently received second courses of megadose therapy at 600–800 mg BCNU/m^2 after their bone marrow had recovered.

Clinical Results

Patients were entered into the study a minimum of 4 months after their initial diagnosis (range 4–18 months). Median survival for the treatment group was 6 months from the date when megadose BCNU-ABMT was administered. 50% of the treatment group (9 of 18) survived longer than 6 months from recurrence.

5 of the 18 patients had significant benefit from therapy, as measured on the basis of clinical status, objective improvement on CT scan, and prolonged survival over 1 year. 2 of these 5 patients (35 and 38 years of age, respectively) had stable disease while on isodose steroids after one course of BCNU (600 mg/m^2). A second course was given at the same dosage 4 and 9 months after the first cycle, which produced significant improvement in clinical status and CT scan results in both patients. These

Table II. Megadose chemotherapy with bone marrow rescue in primary brain tumors

Regimen	ABMT protocol	Responses/ evaluable cases	Duration months	Reference
BCNU 600–1,400 mg/m² Days 1, 2	refrigerated (4 °C) days 3–4	5/18	12–41+	[20, 34]
BCNU 600–2,250 mg/m² Days 1, 2, 3	frozen day 7	4/14	22–44+	[13, 14, 27]
BCNU 1,050–1,200 mg/m², +5-FU Days 1, 2, 3	frozen days 5–7	5/8	12+	[25]
CCNU 390 mg/m² Day 1	frozen days 3–4	0/5	–	[19]
BCNU[1] 800 mg/m² Day 1	frozen or room temperature days 2–3	4/4	1–12+	[5]

[1]Patients treated in clinical remission after surgery with or without irradiation.

patients are currently alive with no evidence of tumor progression 26 and 41 months, respectively, after their first course of therapy.

A 3rd patient, age 36, received BCNU (800 mg/m²) after having had adjuvant CCNU. She improved clinically with no evidence of progression on the CT scans for 20 months. A localized recurrence was treated surgically 23 months after she received megadose chemotherapy and ABMT. She is currently alive with no CT evidence of further recurrence 32 months after the megadose therapy and ABMT.

A 4th patient, age 27, received BCNU (1,000 mg/m²) that resulted in clinical improvement. No change in the tumor bulk was noted on CT scans, but decreased contrast enhancement and mass effect were evident while the patient was on isodose corticosteroids. This patient had received 7,900 rad to the brain as primary therapy at another institution, which makes it difficult to interpret the reason for his long survival; he is currently alive with no evidence of progression 38 months after therapy with megadose BCNU-ABMT.

A 5th patient, age 17, received BCNU (1,200 mg/m²) as a single treatment. A CT scan made 3 months after therapy showed a 30% diminu-

Table III. Toxicity of megadose BCNU with bone marrow rescue[1]

Side effect	BCNU dose range	Number/evaluable cases at same dose	
		Boston [34]	St. Louis, Durham, Cleveland [27]
Neutropenia	nadir ≤500 neutrophils		
	600 mg/m^2	0/3	0/2
	750–1,050 mg/m^2	3/4	7/8
	800 mg/m^2 s/p prior CCNU	6/6	–
	1,200–1,500 mg/m^2	9/9 (1 fatal)	7/7
Thrombocytopenia		nadir ≤20,000	or ≤10,000
	600 mg/m^2	0/3	0/2
	750–1,050 mg/m^2	3/4	5/8
	800 mg/m^2 s/p prior CCNU	6/6	–
	1,200–1,500 mg/m^2	9/9	6/6
Hepatic			
	600–1,050 mg/m^2	2/6	2/19
	≥1,200 mg/m^2	1/2 (fatal)	8/18 (7 fatal)
CNS			
	600–1,200 mg/m^2	2/18 (retinal)	0/19
	≥1,200 mg/m^2	1/2 (fatal)	3/12 (3 fatal)
Pulmonary			
	600–1,050 mg/m^2	1/7 (fatal)	1/15
	1,200 mg/m^2	1/6	4/30 (1 fatal)
	≥1,400 mg/m^2	0/3	1/12

[1] Includes patients with primary brain tumors, melanomas, and other malignant tumors.

tion in tumor bulk. Glioblastoma recurrence and pulmonary interstitial fibrosis developed 1 year after therapy. She succumbed 21 months after receiving megadose BCNU-ABMT.

Table II summarizes clinical data from this study and other megadose nitrosourea-ABMT trials in the treatment of primary brain tumors. A comparison of intertrial data is difficult because of the variable criteria for patient selection, the unknown influence of prior therapy in some patients, and variations in the megadose BCNU and ABMT protocols. A few durable responses have been obtained at several treatment centers using high-dose BCNU with ABMT in patients with recurrent primary brain tumors. Current studies include evaluation of these therapies as an adjuvant after irradiation for newly diagnosed patients with primary brain malignancies [5, 27].

Although only 28% (5 of 18) of the treatment group from the Boston study attained significant responses as measured by prolonged survival, two thirds of patients had evidence of response, as measured by clinical improvement and either stable CT results or diminished CT enhancement of tumor and diminished mass effect [20].

Toxicity

The results regarding toxicity in this study have been reviewed previously [34]. Table III summarizes the major toxicities observed at various BCNU dosage levels in this study and in a recent trial reported by *Phillips* et al. [27] and *Herzig* et al. [18]. The most common serious toxicity was dose-related myelosuppression. About 50% of the treatment group developed fever associated with leukopenia, and 1 patient who received 1,400 mg/m^2 BCNU died of *Clostridia* sepsis associated with leukopenia and diffuse gastrointestinal erosion. 14 patients were given prophylactic platelet transfusions because of thrombocytopenia; however, no bleeding complications were observed.

Severe alterations of mental status developed acutely in 2 patients after they received 1,400 mg/m^2 BCNU. Postmortem examination of 1 of these patients revealed only tumor-related changes. *Schold and Fay* [32] have reported 3 patients with tumors not located in the brain who developed fatal CNS toxicity after receiving BCNU doses of 1,500–2,450 mg/m^2. The autopsies revealed scattered foci of coagulative necrosis in the brain and swollen axis cylinders, changes not detected in patients treated with lesser doses. None of the 10 postmortem examinations from the Boston study revealed these histologic changes; however, only 3 patients received doses greater than 1,200 mg/m^2.

2 patients who received 800 mg/m^2 BCNU developed neurofibrillary retinal infarcts with permanent impairment of vision. Other side effects included nausea (100%), mucositis (<20%), and cutaneous erythema (>75%) associated with peripheral intravenous infusion of BCNU. 1 patient had 50% decreased creatinine clearance after a second dose of BCNU (600 mg/m^2), but an intravenous pyelogram showed no abnormalities and her renal function remains stable. Only 1 patient experienced transient symptoms (nausea, muscle aches, and sternal pressure) associated with marrow infusion.

Conclusion

Megadose chemotherapy with ABMT rescue is an experimental approach to the treatment of refractory malignancies that is being studied in several centers around the world. Preliminary results using high-dose BCNU therapy with ABMT rescue in the treatment of recurrent primary brain tumors indicate that a small group of patients obtain objective response with prolonged disease-free survival. Objective response has also been reported for metastatic tumors in the CNS that are managed using this approach. In the collaborative study from Massachusetts General Hospital and Sidney Farber Cancer Institute, megadose BCNU-ABMT rescue was used in the treatment of 18 patients with recurrent glioblastoma. 5 patients, who received megadose BCNU (600–1,400 mg/m^2) with ABMT rescue, achieved clinical remissions, objective improvement based on CT scan results, and prolonged survival ranging from 21 to 41+ months from the time of tumor recurrence. All responders were under age 40 (average, 31 years). Age also influences the prognosis after standard chemotherapy. The response rate obtained in this study is very similar to recent results from an ongoing trial by *Phillips* and colleagues using megadose BCNU-ABMT in patients with recurrent glioblastomas: responses lasting over 1 year were obtained in 4 of 14 (28.6%) patients in this group [*Phillips,* personal commun.].

Although dose-related myelosuppression was observed after megadose BCNU treatment, despite bone marrow rescue, the major life-threatening toxicity limiting dose escalation was to the liver and the CNS, especially with doses greater than 1,200 mg/m^2. Recent studies in dogs show that equivalent levels of BCNU in brain could be obtained using intracarotid administration without incurring hazardous hematologic or systemic toxicity; this approach potentially could obviate the necessity for bone marrow rescue [manuscript in preparation]. Therapy with intracarotid administration of BCNU is currently under study in clinical trials at the Massachusetts General Hospital and several other institutions.

ABMT is a safe, effective technique that permits escalation of the chemotherapy dose to levels that would otherwise be limited because of hematologic toxicity. This technique may find greater utility in the treatment of recurrent brain tumors using new forms of chemotherapy that presently are limited by their level of marrow toxicity. Further studies in progress will help to define the role for high-dose BCNU and bone marrow rescue as primary therapy of brain tumors after surgery or irradiation.

Acknowledgments

We thank the following physicians who contributed to this study: Drs. *L. Parker, T. Takvorian, G. Canellos, E. Frei* III at the Sidney Farber Cancer Institute and Drs. *S. Lind, A. Weitberg, S. Weitzman, N. Zervas* at the Massachusetts General Hospital. Dr. Fingert is a Clinical Fellow of the American Cancer Society. This work was partly published in detail by *Hochberg* et al. [20] and *Takvorian* et al. [34].

References

1 Abb, J.; Netzel, B.; Rodt, H.V.; Thierfelder, S.: Autologous bone marrow grafts in dogs treated with lethal doses of 1-(2-chloroethyl)-3-cyclohexyl-1-nitrosourea. Cancer Res. *38:* 2157–2159 (1978).

2 Abrams, R.A.; Glaubiger, D.; Simon, R.; Lichter, A.; Deisseroth, A.B.: Haemopoietic recovery in Ewing's sarcoma after intensive combination therapy and autologous marrow infusion. Lancet *ii:* 385–389 (1980).

3 Appelbaum, F.R.; Herzig, G.P.; Ziegler, J.L.; Graw, R.; Levine, A.S.; Deisseroth, A.B.: Successful engraftment of cryopreserved autologous bone marrow in patients with malignant lymphoma. Blood *52:* 85–95 (1978).

4 Blijham, G.; Spitzer, G.; Litam, J.; Zander, A.R.; Verma, D.S.; Vellekoop, L.; Samuels, M.L.; McCredie, K.B.; Dicke, K.A.: The treatment of advanced testicular carcinoma with high dose chemotherapy and autologous marrow support. Eur. J. Cancer *17:* 441–443 (1981).

5 Carella, A.M.; Giordano, D.; Santini, G.; Frassoni, F.; Podesta, M.; Van Lint, M.T.; Bacigalupa, A.; Nati, S.; Vimercati, R.; Occhini, D.; Bistolfi, F.; Lucarelli, G.; Lercari, G.; Marmont, A.M.: High dose BCNU followed by autologous bone marrow infusion in glioblastoma multiforme. Tumori *67:* 473–475 (1981).

6 Cooper, J.S.; Borok, T.L.; Ransohoff, J.; Carella, R.J.: Malignant glioma. J. Am. med. Ass. *248:* 62–65 (1982).

7 Cruz, A.B.; Metter, G.; Armstrong, D.; Aust, J.B.; Fletcher, W.S.; Wilson, W.L.; Richardson, J.: Treatment of advanced malignancy with CCNU (NSC 79037). Cancer *38:* 1069–1076 (1976).

8 Deisseroth, A.; Abrams, R.A.: The role of autologous stem cell reconstitution in intensive therapy for resistant neoplasms. Cancer Treat. Rep. *63:* 461–471 (1979).

9 Dicke, K.A.; Spitzer, G.; Zander, A.R.; Lanzotti, V.J.; Verma, D.S.; Peters, L.J.; Valdivieso, M.; Lotzova, E.; McCredie, K.B.: Autologous bone marrow transplantation in relapsed adult acute leukemia and solid tumors. Transplant. Proc. *11:* 212–214 (1979).

10 Douer, D.; Champlin, R.E.; Ho, W.G.; Sarna, G.P.; Wells, J.H.; Graze, P.R.; Cline, M.J.; Gale, R.P.: High-dose combined-modality therapy and autologous bone marrow transplantation in resistant cancer. Am. J. Med. *71:* 973–976 (1981).

11 Ekert, H.; Ellis, W.M.; Waters, K.D.; Tauro, G.P.: Autologous bone marrow rescue in the treatment of advanced tumors of childhood. Cancer *49:* 603–609 (1982).

12 Epstein, R.B.; Sarpel, S.C.: Autologous bone marrow infusion following high dose

chemotherapy of the canine transmissible venereal tumor (TVT). Exp. Hematol. *8:* 683–689 (1980).

13 Fay, J.W.; Levine, M.N.: Treatment of metastatic melanoma with intensive 1,3-bis(2-chloroethyl)-1-nitrosourea (BCNU) and autologous marrow transplantation (AMTX) (Abstract). Proc. Am. Soc. clin. Oncol. *22:* 532 (1981).

14 Fay, J.W.; Phillips, G.L.: Treatment of neoplastic disease of the central nervous system (CNS) with high-dose 1,3-bis(2-chloroethyl)-1-nitrosourea (BCNU) and autologous marrow transplantation (AMTX) (Abstract). Proc. Am. Soc. clin. Oncol. *21:* 353 (1980).

15 Floersheim, G.L.; Torhorst, J.: Control of a human tumor (Ewing sarcoma) in mice by a single lethal dose of dimethyl-myleran and bone marrow. Int. J. Cancer *27:* 743–748 (1981).

16 Frei, E.; Canellos, G.P.: Dose: a critical factor in cancer chemotherapy. Am. J. Med. *69:* 585–593 (1980).

17 Gratwohl, A.A.; Appelbaum, F.R.; Levine, A.S.; Ziegler, J.L.; Deisseroth, A.B.: Langdauernde, vollständige Remission bei Non-Hodgkin-Lymphomen nach hochdosierter Kombinations-Chemotherapie mit oder ohne autologe Knochenmarktransplantation. Schweiz. med. Wschr. *108:* 1961–1963 (1978).

18 Herzig, G.P.; Phillips, G.L.; Herzig, R.H.; Fay, J.W.; Weiner, R.L.; Wolff, S.; Lazarus, H.M.: High-dose nitrosourea (BCNU) and autologous bone marrow transplantation: a phase I study; in Prestayko, Crooke, Baker, Carter, Schein, Nitrosoureas, pp. 337–341 (Academic Press, New York 1981).

19 Hildebrand, J.; Badjou, R.; Collard-Rouge, E.; Delforge, A.; Marlarme, M.; Spiro, T.; Sztern, B.; Vandensteene, A.; Stryckmans, P.A.: Treatment of brain gliomas with high dose of CCNU and autologous bone marrow transplantation. Biomedicine *32:* 71–75 (1980).

20 Hochberg, F.H.; Parker, L.M.; Takvorian, T.; Canellos, G.P.; Zervas, N.: High-dose BCNU with autologous bone marrow rescue for recurrent glioblastoma multiforme. J. Neurosurg. *48:* 455–460 (1981).

21 Kaizer, H.; Wharam, M.D.; Munoz, L.L.; Johnson, R.J.; Elfenbein, G.J.; Tutschka, P.J.; Braine, H.G.; Santos, G.W.; Leventhal, B.G.: Autologous bone marrow transplantation in the treatment of selected human malignancies: The Johns Hopkins Oncology Center Programm. Exp. Hematol. *7:* suppl. 5, pp. 309–320 (1979).

22 Kornblith, P.L.; Szypko, P.E.: Variations in response of human brain tumors to BCNU in vitro. J. Neurosurg. *48:* 580–586 (1978).

23 Levin, V.A.: Chemotherapy of recurrent brain tumors; in Prestayko, Crooke, Baker, Carter, Schein, Nitrosoureas, pp. 259–267 (Academic Press, New York 1981).

24 Levin, V.A.; Hoffman, W.F.; Pischer, T.L.; Seager, M.L.; Boldrey, E.B.; Wilson, C.B: BCNU-5-fluorouracil combination therapy for recurrent malignant brain tumors. Cancer Treat. Rep. *62:* 2071–2076 (1978).

25 Mortimer, J.; Bay, J.; Livingston, R.B.; Groppe, C.W.: Recurrent primary brain tumors: high dose BCNU and autologous bone marrow 'rescue' (Abstract). Proc. Am. Soc. clin. Oncol. *23:* 174 (1982).

26 Paterson, A.H.G.; English, D.: The effect of BCNU combined with total body irradiation or cyclophosphamide on survival of dogs after autologous marrow grafts. Int. J. Radiat. Oncol. Biol. Phys. *5:* 1585–1587 (1979).

27 Phillips, G.L.; Fay, J.W.; Wolff, S.N.; Karanes, C.; Herzig, R.H.; Herzig, G.P.:

Intensive 1,3-bis(2-chloroethyl)-1-nitrosourea (BCNU) for refractory malignancy: a phase I study (Abstract). Proc. Am. Ass. Cancer Res. *21:* 180 (1980).
28 Rosenblum, M. L.; Dougherty, D. A.; Wilson, C. B.: Rational planning of brain tumor therapy based on laboratory investigation: comparison of single- and multiple-dose BCNU schedules. Br. J. Cancer *41:* suppl. IV, pp. 253–254 (1980).
29 Rosenblum, M. L.; Dougherty, D. A.; Reese, C.; Wilson, C. B.: Potentials and possible pitfalls of human stem cell analysis. Cancer Chemother. Pharmacol. *6:* 227–235 (1981).
30 Sarna, G. P.; Champlin, R.; Wells, J.; Gale, R. P.: Phase I study of high-dose mitomycin with autologous bone marrow support. Cancer Treat. Rep. *66:* 277–282 (1982).
31 Schabel, F. M., Jr.: Nitrosoureas: a review of experimental antitumor activity. Cancer Treat. Rep. *60:* 665–697 (1976).
32 Schold, S. C.; Fay, J. W.: Central nervous system toxicity from high-dose BCNU treatment of systemic cancer. Neurology *30:* 429 (1980).
33 Spitzer, G.; Dicke, K. A.; Verma, D. S.; Zander, A. R.; McCredie, K. B.: High-dose BCNU therapy with autologous bone marrow infusion: preliminary observations. Cancer Treat. Rep. *63:* 1257–1264 (1979).
34 Takvorian, T.; Hochberg, F. H.; Canellos, G.; Parker, L.; Zervas, N.; Frei, E.: The toxicity of high-dose BCNU with autologous marrow support; in Prestayko, Crooke, Baker, Carter, Schein, Nitrosoureas, pp. 155–169 (Academic Press, New York 1981).
35 Thomas, M. R.; Robinson, W. A.; Glode, L. M.; Koeppler, H.; Dantas, M. E.; Morton, N.; Sutherland, J.: Treatment of advanced malignant melanoma with high dose chemotherapy and autologous bone marrow transplantation. Preliminary results, phase I study. Am. J. clin. Oncol. (in press, 1983).
36 Walker, M. D.; Green, S. B.; Byar, D. P.; Alexander, E., Jr.; Batzdorf, U.; Brooks, W. H.; Hunt, W. E.; MacCarty, C. S.; Mahaley, M. S.; Mealey, J., Jr.; Owens, G.; Ransohoff, J.; Robertson, J. T.; Shapiro, W. R.; Smith, K. R.; Wilson, C. B.; Strike, T. A.: Randomized comparisons of radiotherapy and nitrosoureas for the treatment of malignant glioma after surgery. New Engl. J. Med. *303:* 1323–1329 (1980).
37 Weiden, P. L.; Storb, R.; Deeg, H. J.; Graham, T. C.; Thomas, E. D.: Prolonged disease-free survival in dogs with lymphoma after total-body irradiation and autologous marrow transplantation consolidation of combination-chemotherapy induced remissions. Blood *54:* 1039–1049 (1979).
38 Yung, W. A.; Shapiro, J. R.; Shapiro, W. R.: Heterogeneous chemosensitivities of subpopulations of human glioma cells in culture. Cancer Res. *42:* 992–998 (1982).

H. J. Fingert, MD, Hematology-Oncology Unit, Massachusetts General Hospital, Harvard Medical School, Boston, MA 02115 (USA)

Prog. exp. Tumor Res., vol. 28, pp. 79–101 (Karger, Basel 1984)

The Potential of Monoclonal Antibodies as Carriers of Radiation and Drugs for Immunodetection and Therapy of Brain Tumors

M. A. Bourdon[a], *R. E. Coleman*[b], *D. D. Bigner*[a]

Duke University Medical Center, Departments of [a]Pathology and [b]Radiology, Duke University, Durham, N.C., USA

Introduction

The use of antibodies for the treatment of cancer began in the last century with the report of *Hercoult and Richert* in 1895, and was followed by *Ehrlich's* [29] proposal in 1906 that antibodies could be used to deliver chemotherapeutic agents. Subsequently, the studies of *Pressman* et al. [86] and *Bale and Spar* [4] demonstrated the capability of antibodies to bind and alter tumor growth. Although these results have been confirmed over the years [8, 77, 91], there has been only limited use of antibodies for the treatment of cancer in humans [64].

Now, the development of monoclonal antibodies (MABs) defining human tumor-associated antigens is providing solutions to the problems of carrier specificity and quantity, allowing for further progress in the immunodiagnosis and immunotherapy of human cancers, including those of neuroectodermal origin. MABs are being explored as potential tumor targeting carriers of radiation for imaging and radiation therapy, and of a variety of other cytotoxic agents for immunochemotherapy. Among the agents that have been conjugated to antibodies are: the plant toxins ricin, abrin, and gelonin; the bacterial toxins diphtheria toxin-A and *Pseudomonas aeruginosa* toxin; drugs such as chlorambucil, methotrexate (MTX), daunomycin, and neocarzinostatin, as well as enzymes [98] and radionuclides [5, 64]. Each of these cytotoxic agents conjugated to antibodies has been shown to be effective in killing tumor cells in vitro and/or in vivo. Nonetheless, our understanding of the pharmacokinetics and delivery of antibody-drug or radionuclide conjugates to tumors, particularly solid tumors, is incomplete. Although many of the problems, such as those involved with vascular permeability, antigen modulation, and conjugate half-life in the blood and tumor, appear to be resolvable, intratumor cellular heterogeneity may well be the single most difficult hurdle to the development of effective immunochemotherapy. Much work is necessary to elucidate the origins and natural history of tumor heterogeneity – but even this problem may be minimized by detecting tumors early, at a time when tumor heterogeneity is presumably limited, by the use of multiple antibody carriers, each directed at different tumor subsets [37], or by the targeting of basic tumor functions, for example, cell membrane transport mechanisms vital to tumor cell survival [107] and tumor-induced neovascularization [35].

Although the use of immunoconjugates in brain tumor detection and therapy is in its infancy, its success in the diagnosis and treatment of various carcinomas and leukemias is highly encouraging [44, 64, 91]. Before these methods are used clinically, however, their potential and limitations should be evaluated further in experimental animal models in order to maximize the benefits of therapy and minimize adverse effects in humans.

Factors Involved in Conjugate Tumor Localization and Therapy

The effectiveness of any antibody-cytotoxin conjugate depends on its specificity and potency, which in turn are affected by a variety of factors, including antibody affinity, number of antigen binding sites, transport rate, levels of conjugate, cellular heterogeneity and accessibility to the target tissue. Together, these factors define the kinetics of conjugate localization and toxicity. The challenge is to optimize these factors in an effective tumor therapy.

A lack of antibody specificity has been an important factor limiting the development of antibodies as carriers of cytotoxins and radiation. Previously, antibody preparations from antisera have had low levels of specific antibody and large amounts of nonspecific immunoglobulin, and have required extensive absorption of cross-reactive antibody during their preparation. Specific antibodies against tumor antigens rarely have been isolated by immunoabsorption and elution, thus eliminating nonspecific immunoglobulin [25, 63]; but again, the amounts of recovered antibody were low. The development of the hybridoma methodology now makes it possible to obtain, in virtually unlimited supply, MABs with single determinant specificity. MABs have been developed against a variety of human neuroectodermal tumor-associated antigens [10, 14, 17, 27, 54, 94, 97, 99, 110, 111]. Although tumor-specific antigens are still elusive, oncofetal and differentiation antigens may well be adequately specific targets. Oncofetal and developmental antigens, such as the neuroectodermal oncofetal antigens defined by MABs [110], can serve as operationally tumor-specific antigens because they are not expressed in normal adult tissues. Differentiation antigens may also prove useful, as has been shown for the common acute lymphoblastic leukemia antigen (CALLA) expressed by T-cell leukemias and a subpopulation of marrow cells, and for B-cell immunoglobulin idiotype, expressed by a B-cell clone and the B-cell leukemia derived from it [72, 90]. Conjugates directed against these antigens are deleterious to only a small population of normal cells that presumably can be regenerated by progenitor cell populations not expressing the differentiation antigen.

Neuroectodermal tumor-differentiation or organ-associated antigens defined by MAB may also prove to be useful in tumor immunodiagnosis and immunochemotherapy. Neuroectodermal lymphoid-associated antigens have been defined by studies with MABs raised against human neuroblastoma cell lines [54] and fetal tissue [111]. Neuroectodermal mesenchymal-associated antigens expressed by human neuroblastomas and glioblastomas have also been defined by studies with MABs [14, 97]. Other MABs reactive with melanomas [27], astrocytomas [17] and glioblastomas [10] define either neuroectodermal tumor subsets or more broadly expressed neuroectodermal tumor-associated antigens.

Several strategies may operationally increase antibody specificity. If site activation of conjugates is possible, for example with antibody-^{10}B activation with slow neutrons, then a broader tumor-tissue reactive antibody could be rendered tumor-specific. Localization of ^{10}B-labeled MAB in other tissues would have little consequence, as activation of ^{10}B would occur only in the tumor tissue exposed to slow neutrons. Specificity may also be enhanced in solid tumors by local administration of conjugates into the proximal blood supply of the tumor, for example by intracarotid injection in the case of intracranial tumors or by intratumoral injection. By allowing the conjugate to 'see' the tumor-expressed antigen first, it may be possible to reduce the levels of tissue binding at other sites.

The kinetics of conjugate binding and cytotoxicity are also important in determining the value of a particular conjugate in localization and therapy studies. Differences in antibody affinity or number of antigen binding sites can greatly affect the rate of conjugate binding and toxicity as well as the total amount of bound conjugate [80]. The implication for therapy is that changes in affinity, receptor number, transport rate, or relative conjugate concentration could have great impact in terms of effective tumor cell kill. To achieve a specific level of tumor cell kill, an immunotoxin of lower affinity would require much longer exposure to the target cells or higher concentrations. Similarly, cells with a reduced receptor number would require longer exposure to a given immunotoxin preparation than would cells with higher numbers of receptors, or they would escape therapy entirely. The situation is further complicated when cellular receptor heterogeneity, which is likely to be characteristic in most tumors, is considered. Receptor heterogeneity may reflect both quantitative and absolute differences in receptors in different subpopulations of the tumor [1, 9], and it could result in effective therapy of only a fraction of the tumor mass.

Antigen shedding and modulation can also affect conjugate localization and toxicity. The presence of soluble shed [101], or secreted antigens, such as carcinoembryonic antigen [64], could result in absorption and elimination of conjugate, thus reducing or abrogating antibody binding of conjugate to the target cells or tissue. Modulation of surface antigen by target cells is another mechanism by which tumor cells might escape antibody binding and effective immunotherapy. By internalizing cell-bound antibody-antigen complex, cells can appear antigen-negative, thus preventing further antibody binding. This has been shown to occur in at least one antigen system in which CALLA-positive leukemia cells escaped MAB J5-mediated cytolysis by antigen modulation [91]. Antigen modulation may be less of a problem when dealing with immunotoxins that must be internalized to be cytotoxic; however, this may depend on the potency of the toxic agent. In the case of antibody-ricin conjugates, a single A chain may suffice to kill the target cell; however, drugs such as MTX or daunomycin may not be transported in sufficient amounts to be cytotoxic. Enzymatic degradation of cell-bound antibody by cell surface or cell-secreted enzymes may be yet another mechanism by which tumor cells escape conjugate cytotoxicity [36].

The influence of accessibility of antibodies and their drug or radionuclide conjugates to tumor tissues is an important factor that has been debated but seldom examined in studies of radiolabeled antibody imaging and conjugate therapy. Accessibility is particularly important when dealing with intracranial tumors because the blood-brain barrier (BBB) may exclude antibody or antibody conjugates from the brain parenchyma. As the development of MAB radioimaging and drug delivery progresses, it is increasingly important to define the physiologic factors effecting accessibility of antibody to the tumor. In solid tumors, three major factors generally would be expected to affect access to the tumor by antibodies or other agents: (a) the extent of vascularization, (b) the dynamics of blood flow and extracellular fluid circulation, and (c) vascular permeability.

The evidence from ultrastructural [81] and permeability [45] studies indicates that much of the vasculature of experimental and human brain tumors, particularly glioblastoma, does not have an intact BBB. Important questions remain about the extent of BBB disruption at the margins of tumors and whether the heterogeneity of vascular permeability seen in several animal models might allow portions of the tumor to escape detection and therapy [11, 45].

Both low blood flow rates and limited diffusion into and out of the tumor tissue could seriously limit the delivery of antibody-drug conjugates. The studies of *Groothuis* et al. [vol. 27, pp. 132–153] have begun to define the parameters of blood flow and permeability in experimental brain tumors and should provide valuable methodology for the evaluation of antibody-drug conjugate pharmacokinetics in animal models and humans. Osmotic BBB disruption may provide an additional means of reaching those areas of a brain tumor in which the BBB is intact. The work of *Neuwelt* et al. [79] and *Hicks* et al. [47] has demonstrated the utility of this procedure in both animal models and humans in enhancing delivery of drug or antibody to brain tissues.

Since antibodies have begun to be used for either detection or therapy of tumors in humans, the problems of antibody clearance and immunogenicity have become more important. The intact immunoglobulin antibody, particularly the Fc portion, is highly immunogenic [100]. Repeated injection of xenoimmunoglobulin would be expected to elicit host antibodies against the foreign protein. Such an antibody response could pose a serious threat to the health of the host, as well as blocking target cell binding by the injected antibody. It is surprising, therefore, that the development of host responses to xenoimmunoglobulin has not been a serious problem in humans [91, 95]. It may be possible to further limit host immune responses to xenoimmunoglobulin through the use of immunosuppressive drugs [76]. However, it would seem prudent to explore the use of the much less immunogenic antibody Fab' and $F(ab')_2$ antigen-binding fragments [100].

The use of Fab' or $F(ab')_2$ fragments would be expected to reduce nonspecific antibody binding as a result of cellular Fc receptor binding by intact immunoglobulin. The Fab' and $F(ab')_2$ antibodies also equilibrate rapidly with the extracellular space and have rapid clearance from the circulation [31, 100]. An antibody $F(ab')_2$-drug conjugate, therefore, would have the advantages of rapid diffusion to the extracellular space surrounding the tumor cells in solid tumors and rapid reduction in normal tissue concentrations. These properties would enhance tumor detection by radiolabeled antibody imaging and would reduce nonspecific host toxicity. In the case of radionuclide antibody therapy, rapid systemic clearance could reduce nonspecific radiation toxicity [30]. It may ultimately be possible to further reduce antibody antigenicity by the use of human MABs [99].

Monoclonal Antibodies as Carriers of Plant and Bacterial Toxins

Abrin and Ricin

Abrin and ricin have served as model agents for the study of targeted therapy. The separation of specific binding function and effector function in different peptide chains is the prototype of all immunotoxin conjugates [34, 61, 83]. Although the results of studies with this model have been encouraging, the level of tumor growth suppression varied greatly among the various tumors examined. In addition, direct abrin or ricin treatment has required the use of high sublethal doses that could cause damage to normal host organs [83]. Work also has been done to target specifically either whole abrin or ricin toxin or the toxic A chain alone [12, 18, 41, 82, 88, 89, 106, 113, 114].

Antibodies, particularly MABs, can have exquisite specificity that should allow specific delivery of toxin to target cells. Attempts to utilize this specificity for delivery to neoplasms have taken three approaches: (a) the use of whole toxin conjugated to antibody; (b) the conjugation of A chain with antibody, and (c) the use of hybrid antibodies with dual specificity for the target antigen and toxin. MAB-ricin conjugates appear to be effective cytotoxic agents [49, 114], although a serious limitation to their application in humans is the need to block nonspecific ricin toxicity with galactose.

The cytotoxicity of antibody conjugates with ricin A chain has been examined in a variety of animal and human tumor systems [12, 43, 56, 77, 89]. Antibody A-chain conjugates provide specific binding of toxin; however, reduced nonspecific toxicity and increased cell specificity are generally mitigated by reduced toxicity of the conjugate as compared to the intact toxin – although it may be possible to increase A-chain toxicity by adding free B chain [115]. In vitro studies [12] with MAB anti-Thy-1 A-chain conjugate showed specific toxicity at concentrations 2,000- to 6,000-fold lower than A chain alone. Antibody A-chain conjugates prolong survival significantly in mice bearing small tumors, but not in mice with large tumors [12, 57], suggesting that it may be necessary to reduce the tumor mass by conventional surgery and radiation therapy before treating the remaining tumor by immunochemotherapy.

The use of affinity-purified antibody and MAB defining human tumor-associated antigen conjugated to ricin A chain has also been investigated [43, 56, 88, 96, 107]. Monoclonal anticolorectal carcinoma-associated antigen conjugated to ricin or diphtheria A chain selectively killed, in vitro, appropriate colorectal cell lines, but not an antigen-negative melanoma cell line [43]. Cells of the human breast carcinoma cell line ZR75-1 have been specifically killed by a conjugate of A chain and MAB H59 [56]. A very well-controlled study by *Raso* et al. [89], investigating the toxicity of MAB to CALLA conjugated to ricin A chain directed against CALLA-positive NALM-1 cells, showed that the conjugate was effective at 2,000-fold lower concentrations than was A chain alone. Comparison of Fab' and $F(ab')_2$ MAB fragments in targeting ricin A chain showed that the $F(ab')_2$ conjugate was 70-fold more effective than the Fab' conjugate; valency may be a factor in this effect [68]. MABs to neuroectodermal antigens are also being examined as carriers of ricin A chain. *Seeger* [96] has reported preliminary results showing in vitro toxicity of conjugates of ricin A chain and MAB 390, which defines human Thy-1 expressed by neuroblastomas and gliomas. MABs currently are being used to detect bone marrow metastases of neuroblastoma [16].

Although the use of abrin in antibody-toxin conjugates has been much more limited than that of ricin, the results parallel those with ricin conjugates [105, 106].

Gelonin and Phytolacca americana *Toxin*

Gelonin and *Phytolacca americana* (pokeweed) toxin, two plant lectins that have the same ribosomal inactivating activity as ricin or abrin A chain but lack a B-chain cell receptor binding site, have a potential advantage over ricin and abrin, in that they lack cellular toxicity in the absence of conjugation to cell-specific antibodies [102]. In vitro experiments with monoclonal anti-Thy-1.1-gelonin [104] showed high levels of specific inhibition of protein synthesis of concanavalin A-stimulated AKR splenocytes, BW5147 lymphoma cells, and AKR-A lymphoma cells. In vitro experiments [69] with a Fab' anti-L1210 mouse leukemia cell line conjugated to *Phytolacca americana* toxin showed identical toxicity toward L1210 cells by both the Fab'-*Phytolacca americana* toxin conjugate and Fab'-A-chain conjugate.

Diphtheria Toxin and Pseudomonas aeruginosa *Toxin*

Diphtheria toxin (62,000 m.w.) and *Pseudomonas aeruginosa* toxin (66,000–72,000 m.w.) are bacterial toxins [84]. Structurally, they are continuous polypeptide chains with two functional domains that, after protease nicking and thiol reduction, are separated into fragments A and B, which are functionally equivalent to those of the plant toxins abrin and ricin. The bacterial toxin A fragment differs from ricin A chain in enzymatically inactivating eukaryotic ribosomal elongation factor EF-2 [112], a single molecule of A chain being sufficient to halt all protein synthesis within a cell.

The selective toxicity of diphtheria toxin conjugates has been examined in several studies [19, 41, 42, 76, 77, 92, 106, 107]. *Ross* et al. [92] and *Thorpe* et al. [106] have been highly successful in inhibiting protein synthesis and growth of human lymphoblastoid cell lines with diphtheria toxin – IgG antihuman lymphocyte globulin, which they found to be 1,000- to 2,000-fold more toxic than diphtheria toxin alone. Similar results were obtained by *Gilliland and Collier* [42] in a model system in which antibody to concanavalin A was used to target diphtheria toxin A fragment to concanavalin A-coated mouse and hamster cell lines.

Antibody-diphtheria toxin conjugates also prolong survival and reduce tumor incidence in mice bearing DNP-haptenated or SV-40-infected tumor cells using appropriate specific antibodies [74, 75]. In two recent studies, MABs to human transferrin receptor or human colorectal carcinoma cell lines were conjugated to diphtheria toxin A fragment. Monoclonal antitransferrin receptor coupled to diphtheria toxin A fragment had toxic effects on CCRF-CEM human T leukemia cells but not BW 5147 mouse lymphoma [107]. Similarly, human colorectal carcinoma cell lines treated in vitro with monoclonal anticolorectal antibody conjugated to diphtheria toxin A fragment produced virtually complete inhibition of protein synthesis and reduction of plating efficiency of treated cells to 0–2% [43].

There are several disadvantages to the use of diphtheria or *Pseudomonas* toxins that may well discourage further study of these toxins in antibody-toxin conjugates. A technical disadvantage is the difficulty in obtaining these toxins, particularly purified A fragment, as compared to the plant toxins. An important disadvantage of diphtheria toxin in particular is that a large proportion of the population is actively immunized against it. This could significantly reduce its effectiveness by its binding to circulating neutralizing antibody, and could complicate therapy and patient management if immunosuppression became necessary.

Monoclonal Antibodies as Carriers of Chemotherapeutic Agents

Chemotherapy has proven efficacy in the treatment of human cancers, but the drugs that are now in use are not tumor-specific. Their effectiveness depends on altering or inhibiting basic protein or nucleic acid pathways of proliferating cells, which includes not only the tumor but hemopoietic stem cells, gastrointestinal epithelia, bladder epithelia, and oral mucosa as well.

If the effective drugs could be targeted specifically to tumor cells, then it might be possible to maintain therapeutic levels at the tumor site over an extended time and with reduced systemic toxicity. This goal has been achieved in a number of in vitro and in vivo studies in which drugs have been coupled covalently or noncovalently to molecules that have target cell specificity. Antibodies, including MABs with tumor or cell type specificity, have been coupled to a variety of chemotherapeutic agents, including *p*-phenylenediamine mustard [21], melphalan [48], triaziquinone [40], and vindesine [52]. The most encouraging results have been found in studies with MTX, chlorambucil, daunomycin, adriamycin, alpha-amanitin, and neocarzinostatin.

Methotrexate

Mathé et al. [70], in 1958, were the first to investigate the use of antibodies for drug delivery. Diazo-derivatized MTX was coupled to an immunoglobulin fraction prepared from antisera raised against the mouse leukemia cell line L1210. This conjugate was effective in treating L1210 tumors transplanted in mice. Antibody-MTX conjugates also prolonged the survival of mice bearing ovarian carcinoma [15]; and MABs to two murine histocompatibility antigens were shown to specifically target MTX-containing liposomes to blast cells of CBA mice; nonspecific drug toxicity was not detected [59]. The use of MAB to specifically target drug-containing liposomes could provide a new approach to drug delivery, particularly with drugs that have limited solubility or limited ability to cross cell membranes.

Chlorambucil

The therapeutic effects of antibody-chlorambucil conjugates have been examined in vitro and in vivo in animal tumor models [26, 28, 33, 38, 48] and, in at least one instance, in humans [38]. The early studies of *Ghose* et al. [38] and *Flechner* [33] showed that anti-Ehrlich ascites cell antibodies in complex with chlorambucil were effective against Ehrlich ascites tumor cells in vitro and in mice. *Dullens* et al. [28] have reported the effectiveness of antibody-chlorambucil conjugates on two murine tumor cell lines, DBA/2 SL2 lymphoma and HPM melanoma, that differ greatly in their tumor growth characteristics and response to antibody-chlorambucil conjugates. SL2 lymphoma grows rapidly and metastasizes readily, and its growth in mice is not inhibited by antibody-chlorambucil conjugate. HPM melanoma grows slowly as a solid tumor, and when transplanted in mice, its growth is inhibited by antibody-chlorambucil conjugate. While the precise basis for the varied susceptibility of these tumors to antibody-drug treatment is undefined, the results nevertheless suggest that characteristics such as the rate of tumor growth and metastatic potential may have important implications for antibody-drug therapy.

Daunomycin, Adriamycin

Daunomycin and adriamycin, closely related antibiotics with antitumor activity, have been conjugated to a variety of antitumor antibodies [7, 50, 51, 60, 85, 87, 108]. Daunomycin conjugated to anti-alpha-fetoprotein is specifically cytotoxic for both a murine and a rat hepatoma cell line in vitro [7, 108] and for rat hepatoma xenografts [108]; antibody-daunomycin conjugates are effective in treating murine lymphoid tumors in mice [51, 60]. A MAB-adriamycin conjugate effective against rat mammary carcinoma cells in vivo has been described by *Pimm* et al. [85]. This well-documented study showed that MAB to Sp4 cells specifically localized in tumor tissues and, when conjugated to adriamycin, significantly delayed tumor growth and prolonged survival in Sp4-bearing rats. Conjugate treatment beginning 6 days after tumor inoculum resulted in significantly prolonged survival with 1/5th the adriamycin dose necessary for an effect on tumor growth. This study is highly encouraging, as it clearly demonstrates that a MAB-drug conjugate can have the hoped-for advantages of highly selective tumor localization and enhanced drug toxicity.

Alpha Amanitin

Enhancement of drug toxicity and specificity by drug conjugation with MAB has also been reported using monoclonal anti-Thy-1.2 coupled to an alpha amanitin azo derivative [22]. Alpha amanitin is a highly toxic inhibitor of RNA polymerase II [20]. In vitro test results show that the antibody-drug conjugate is 133–277 times more toxic to Thy-1.2$^+$ S49.1 lymphoma cells than is alpha amanitin, and is up to 700 times more toxic than azo-amanitin. While alpha amanitin is far less potent than ricin, for example, the toxicity of their antibody conjugates compare favorably on a molar basis.

Neocarzinostatin

Neocarzinostatin is an antibiotic that potentially may be a useful toxic agent in the development of effective antibody-drug conjugates. It is small (10,700 m.w.) and can mediate cellular toxicity either intracellularly or on the cell surface [58, 93]. *Raso and Griffin* [87] have reported that anti-human lymphocyte immunoglobulin-neocarzinostatin was effective in killing cells of the human lymphoid cell line SB. Recently, *Jung* et al. [53] reported on the conjugation of MABs to neocarzinostatin, opening the way for evaluation of these conjugates as specific cytotoxic agents.

Monoclonal Antibodies as Carriers for Radionuclides in Imaging and Therapy

Radionuclides

Radionuclides offer several advantages over the agents we have discussed, but have many of their disadvantages also. The activity of radiolabeled compounds can readily be determined in vitro and, more important, in vivo. Through the use of radioscintigraphic imaging and quantitative autoradiographic techniques, distribution, kinetics, and dosimetry can be studied directly [11, 64]. In addition, radionuclides produce their tumoricidal effects without entering the tumor cell. Furthermore, the effects of ionizing radiation produced by radionuclides is not confined to the cell, but can encompass an area many times that of a single cell diameter. This is both an advantage and a disadvantage, as radionuclides kill indiscriminately and continuously, exposing normal tissues and blood as well as the tumor to ionizing radiation [30]. The problem in therapy is to target as much of the radionuclide to the tumor as fast as possible while keeping the exposure of normal tissue to a minimum. This is less of a problem in diagnostic procedures, as they require much lower levels of radionuclide.

The need for specific targeting of radionuclides has led to the examination of antibodies, particularly MABs, as carriers, and to studies of their potential for detection and therapy of tumors [39, 44, 55, 64]. Radiolabeled MABs are proving especially useful in antibody localization studies, in work involving radioimaging of human tumor xenografts [6, 13, 78, 103], and in clinical studies [32, 65]. The encouraging results of these investigations suggest that radioimmunotherapy may now be possible.

Radioimmunodetection and Therapy of Brain Tumors

Following the early studies in animal tumor models [5, 23, 86], *McCardle* et al. [71] and *Marrack* et al. [67] evaluated antifibrinogen localization using radiolabeled antifibrinogen antibodies in a large number of patients with brain tumor. In these studies, approximately 80% of brain tumors, both primary and metastatic, were detected by ^{131}I-antifibrinogen radioscintigraphy. Although these results were promising, the nonspecific nature, variable expression, and heterogeneous distribution of the fibrinogen antigen in brain tumors greatly limited its utility in radioimaging and therapy.

In an attempt to localize radiolabeled antibodies more specifically in human glioblastoma, *Day* et al. [24] and *Mahaley* et al. [66] used immunoabsorbent purified ^{131}I-labeled antiglioblastoma antibody preparations that localized within tumors in 11 of 12 cases, as determined by radioscintigraphy before surgery and tissue gamma counting and autoradiography of the surgically removed tissue. However, the difficulties in obtaining sufficient quantities of specific antibody from heteroantisera made further application of radiolabeled antiglioma antibody impractical for brain tumor imaging or treatment.

With the discovery of MABs, it may now be possible to target

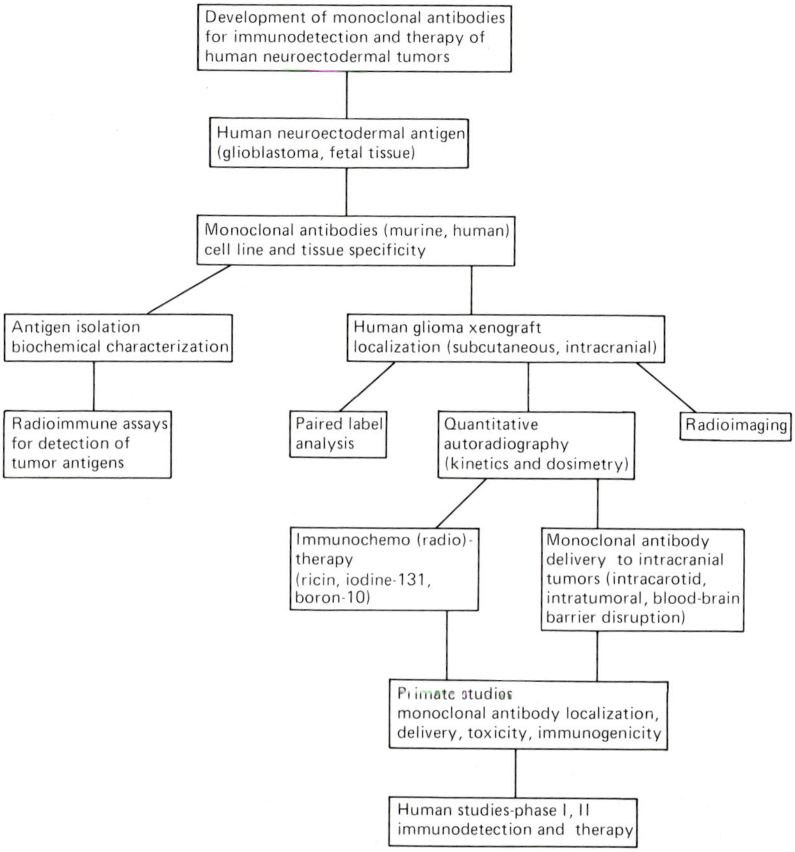

Fig. 1. An outline for the development of MAB immunodetection and immunochemotherapy of human brain tumors.

radionuclides specifically to human brain tumors in diagnostic and therapeutic amounts. To evaluate the potential of MABs as carriers of radioactive molecules in human glioblastoma, we have undertaken a stepwise program of studies, beginning with the development and characterization of MABs directed against glioma-associated antigens, followed by localization, imaging, and therapy studies in the human glioma nude mouse xenograft model, and ultimately by imaging and therapy studies in humans (fig. 1). As the route of radiolabeled-antibody delivery may play an important role in the extent to which labeled antibody ultimately can be localized in human gliomas [25], intravenous, intracarotid, and intratumoral routes are being evaluated in intracranial tumor models. Positive results in these studies may allow for the rational development of protocols for phase I studies in glioblastoma patients.

Monoclonal Antibody Characterization and Immunolocalization Studies

We recently reported our findings on the characterization of 8 MABs defining glioma-associated antigens [10, 14, 110, 111]. Our MABs are directed against cell surface or matrix antigens with highly restricted distribution in vitro and in normal and neoplastic tissues. None are reactive with normal brain. One of these MABs, 81C6, defines a glioma-mesenchymal extracellular matrix antigen present in the basement membranes of hyperplastic and glomeruloid capillaries of human glioblastoma, glioblastoma cell lines, and glioma xenografts in nude mice [14]. Immunochemical and immunohistologic results indicate that the glioma-mesenchymal extracellular matrix antigen is distinct from normal fibronectin, collagen, and proteoglycan.

In our localization studies, we examined radiolabeled 81C6 localization by paired label analysis, radioscintigraphy, and autoradiography in tumors from the human glioma cell lines, U-251 MG and D-54 MG, and the human breast carcinoma, DU-4475.

Paired label experiments, with ^{125}I-labeled 81C6 and ^{131}I-labeled myeloma immunoglobulin 45.6 as control immunoglobulin of the same IgG_{2b} isotype, demonstrated high levels of specific localization of 81C6 in U-251 MG and D-54 MG glioma xenografts, but not in DU-4475 breast carcinoma xenografts [86]. The upper range of the localization index [78] in the glioma xenografts was as high as 18, which compares favorably with the results of radiolabeled MAB localization in human teratocarcinoma xenografts in nude mice reported by *Moshakis* et al. [78]. Nearly 14.3% of the initial injected dose of ^{125}I-81C6 was present in D-54 MG tumors on day 5. Ratios of MAB 81C6 tumor to normal tissue in D-54 MG tumor-bearing mice ranged from an average of 15 in kidney to 106 in brain. The very large differences between tumor and brain localization could prove to be important in the results of therapy of intracranial tumor xenografts in either mice or rats. It is possible that reducing the half-life of labeled MAB by using $F(ab')_2$ or Fab' antibody fragments would increase further the tumor-to-tissue antibody ratio.

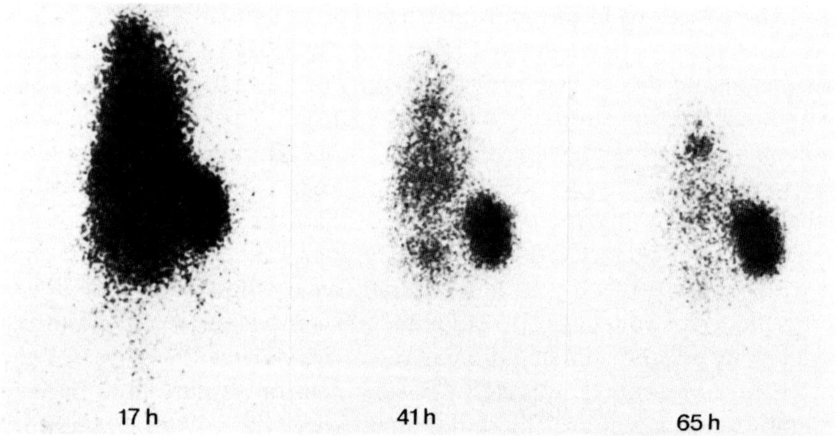

Fig. 2. Radioscintigraphy of athymic mice bearing U-251 MG flank tumor following injection of ^{131}I-labeled MAB 81C6. Tumor bearing mice were injected with radiolabeled MAB 81C6 and radioimaged at the times indicated.

As a direct consequence of labeled-81C6 localization in glioma xenografts, it has been possible to image the glioma xenograft of tumor-bearing mice. Radioscintigraphy of ^{131}I-labeled 81C6 injected into U-251 MG tumor-bearing mice showed increased ^{131}I-81C6 uptake in the flank tumor at 20 h and clear imaging of the tumor by 48–60 h (fig. 2); computer-assisted radioscintigraphy of tumors allows us to follow and quantitate tumor size and radiation dose during the course of therapy studies with ^{131}I-81C6. Autoradiographs of U-251 MG tumors injected with radiolabeled 81C6 demonstrated localization of 81C6 in high concentration within the stroma of the tumor xenograft, whereas control immunoglobulin was distributed in a nonspecific manner throughout the tumor.

Monitoring of labeled antibody preparations as well as labeled antibody in the plasma of injected tumor-bearing mice by gel chromatography and binding assays has shown that both the antibody specificity and structure are not altered detectably by radiolabeling or injection. Another indication that labeled 81C6 is not undergoing any unusual degradation or aggregation is reflected in the half-life of both 81C6 and nonspecific IgG 45.6 (both IgG_{2b}) in the blood of control and tumor-bearing mice, which compares well with the half-life of various mouse myeloma IgG_{2b} reported by others [31]. We have also shown that radiolabeled antibody eluted from U-251 MG tumors retains antibody binding activity and specificity.

The results of localization and imaging studies with human glioma xenografts and radiolabeled 81C6 are very encouraging. In light of these results, an old theory may provide exciting new results: it has been proposed that the activation of ^{10}B by slow neutrons, which results in a recoil lithium atom and an alpha particle, could be an effective means of delivering ionizing radiation locally in tissues [3, 62]. The advantage of a ^{10}B-labeled MAB would be the combined characteristics of high target-specificity and site-activation of ^{10}B. The long experience with ^{10}B slow neutron capture therapy of human brain tumors [2], the availability of procedures for coupling ^{10}B compounds to antibodies [46, 73], and the availability of antineuroectodermal tumor MABs set the stage for the evaluation of antibody-targeted ^{10}B slow neutron capture as a form of therapy of brain tumors. The studies in progress will provide a framework within which to pursue the possibilities of therapy of tumors of the central nervous system with other MABs directed against neuroectodermal tumor-associated antigens in humans. In our studies, we are attempting to define the problems of antibody localization, radiation dosimetry, and toxicity in rodent and subsequently primate models of human gliomas, in order to begin the development of rational and safe phase-I and phase-II imaging and treatment studies in humans (fig. 1).

References

1 Albino, A.P.; Lloyd, K.O.; Houghton, A.N.; Oettgen, H.F.; Old, L.J.: Heterogeneity in surface antigen and glycoprotein expression of cell lines derived from different melanoma metastases of the same patient. Implications for the study of tumor antigens. J. exp. Med. *154:* 1764–1778 (1981).

2 Asbury, A.K.; Ojemann, R.G.; Nielson, S.L.; Sweet, W.H.: Neuropathologic study of fourteen cases of malignant brain tumor treated by boron-10 slow neutron capture radiation. J. Neuropath. exp. Neurol. *31:* 278–303 (1972).

3 Bale, W.F.: Relative effects of alpha, beta, gamma and neutron radiation. Proc. natn. Cancer Conf. *2:* 967–976 (1952).

4 Bale, W.F.; Spar, I.L.: Studies directed toward the use of antibodies as carriers of radioactivity for therapy. Adv. biol. med. Phys. *5:* 286–356 (1957).

5 Bale, W.F.; Spar, I.L.; Goodland, R.L.: Experimental radiation therapy of tumors with ^{131}I-carrying antibodies to fibrin. Cancer Res. *20:* 1488–1494 (1960).

6 Ballou, B.; Levine, G.; Hakala, T.; Solter, D.: Tumor location detected with radioactively labeled monoclonal antibody and external scintigraphy. Science *206:* 844–846 (1979).

7 Belles-Isles, M.; Page, M.: In vitro activity of daunomycin-anti-alphafetoprotein conjugate on mouse hepatoma cells. Br. J. Cancer *41:* 841–842 (1980).

8 Bernstein, I.D.; Nowinski, R.C.: Monoclonal antibody treatment of transplanted and spontaneous murine leukemia. Prog. Cancer Res. Ther. *21:* 97–111 (1982).

9 Bigner, D.D.; Bigner, S.H.; Ponten, J.; Westermark, B.; Mahaley, M.S.; Ruoslahti, E.; Herschman, H.; Eng, L.F.; Wikstrand, C.J.: Heterogeneity of genotypic and phenotypic characteristics of 15 permanent cell lines derived from human gliomas. J. Neuropath. exp. Neurol. *40:* 201–229 (1981).

10 Bigner, D.D.; Wikstrand, C.J.: Characterization of three monoclonal antibodies (MCA) raised against the human glioma cell line D-54 MG (manuscript submitted, 1982).

11 Blasberg, R.G.; Kobayashi, T.; Patlak, C.S.; Shinohara, M.; Miyoaka, M.; Rice, J.M.; Shapiro, W.R.: Regional blood flow, capillary permeability, and glucose utilization in two brain tumor models: preliminary observations and pharmacokinetic implications. Cancer Treat. Rep. *65:* 3–12 (1981).

12 Blythman, H.E.; Casellas, P.; Gros, O.; Gros, P.; Jansen, F.K.; Paolucci, F.; Pau, B.; Vidal, H.: Immunotoxins: hybrid molecules of monoclonal antibodies and a toxin subunit specifically kill tumor cells. Nature, Lond. *290:* 145–146 (1981).

13 Bourdon, M.A.; Wikstrand, C.J.; Coleman, R.E.; Bigner, D.D.: Specificity and tumor localization of a monoclonal antibody defining a human glioma associated extracellular matrix antigen. J. Neuropath. exp. Neurol. *41:* 367 (1982).

14 Bourdon, M.A.; Wikstrand, C.J.; Bigner, D.D.: A human glioma-mesenchymal extracellular matrix antigen defined by monoclonal antibody. Cancer Res. *43:* 2796–2805 (1983).

15 Burstein, S.; Knapp, R.: Chemotherapy of murine ovarian carcinoma by methotrexate-antibody conjugates. J. medicinal Chem. *20:* 950–952 (1977).

16 Byrd, R.L.; Jonak, Z.L.; Kennett, R.H.: Monoclonal antibodies against neuroblastomas and leukemias: detection of metastases and of molecules involved in growth regulation. Prog. Cancer Res. Ther. *21:* 219–229 (1982).

17 Cairncross, J.G.; Mattes, M.J.; Beresford, H.R.; Albino, A.P.; Houghton, A.N.; Lloyd, K.O.; Old, L.J.: Cell surface antigens of human astrocytoma defined by mouse monoclonal antibodies: identification of astrocytoma subsets. Proc. natn. Acad. Sci. USA (in press, 1982).

18 Cawley, D.B.; Herschman, H.R.; Gilliland, D.G.; Collier, R.J.: Epidermal growth factor-toxin A chain conjugates: EGF-ricin A chain is a potent toxin while EGF-diphtheria fragment A is non-toxic. Cell *22:* 563–570 (1980).

19 Chang, T.M.; Dazord, A.; Neville, D.M.: Artificial hybrid protein containing a toxic protein fragment and a cell membrane receptor-binding moiety in a disulfide complex. II. Biochemical and biological properties of diphtheria toxin fragment A-S-S-human placental lactogen. J. biol. Chem. *252:* 1515–1522 (1977).

20 Cochet-Meilhae, M.; Chambon, P.: Animal DNA-dependent RNA polymerase. II. Mechanism of the inhibition of RNA polymerases B by amatoxins. Biochim. biophys. Acta *353:* 160–184 (1974).

21 Davies, D.A.L.; O'Neill, G.J.: Specific cancer therapy by drugs attached to tumor-specific antibodies. Ann. N.Y. Acad. Sci. *277:* 670 (1976).

22 Davis, M.T.; Preston, J.F.: A conjugate of alpha-amanitin and monoclonal immunoglobulin G to Thy 1.2 antigen is selectively toxic to T lymphoma cells. Science *213:* 1385–1388 (1981).

23 Day, E.D.; Planinsek, J.; Korngold, L.; Pressman, D.: Tumor-localizating antibodies

purified from antisera against murphy rat lymphosarcoma. J. natn. Cancer Inst. *17:* 517–532 (1956).

24 Day, E.D.; Mahaley, M.S.; Woodhall, B.; Pireher, F.: Localization of purified ^{125}I-anti-tumor radioantibodies in human brain tumors. J. nucl. Med. *5:* 357–358 (1964).

25 Day, E.D.; Lassiter, S.; Woodhall, B.; Mahaley, J.L.; Mahaley, M.S.: The localization of radioantibodies in human brain tumors. I. Preliminary exploration. Cancer Res. *25:* 773–778 (1965).

26 DeWeger, R.A.; Dullens, H.F.J.; Den Otter, W.: Eradication of murine lymphoma and melanoma cells by chlorambucil-antibody complexes. Immunol. Rev. *62:* 29–45 (1982).

27 Dippold, W.G.; Lloyd, K.O.; Li, L.T.C.; Ikeda, H.; Oettgen, H.F.; Old, L.J.: Cell surface antigens of human malignant melanoma: definition of six antigenic systems with mouse monoclonal antibodies. Proc. natn. Acad. Sci. USA *77:* 6114–6118, 1980.

28 Dullens, H.F.J.; Vennegoor, D.; DeWeger, R.A.; Woutersen, F.; Woutersen, R.A.; Den Otter, W.: Comparison of various forms of therapies in two different mouse tumor systems. Cancer Treat. Rep. *63:* 99–109 (1979).

29 Ehrlich, P.: Collected studies on immunity, p. 441 (Wiley & Sons, New York 1906).

30 Ettinger, D.S.; Order, S.E.; Wharam, M.D.; Parker, M.K.; Klein, J.L.; Leichner, P.K.: Phase I–II study of isotopic immunoglobulin therapy for primary liver cancer. Cancer Treat. Rep. *66:* 289–297 (1982).

31 Fahey, J.L.; Sell, S.: The immunoglobulin of mice v. the metabolic (catabolic) properties of five immunoglobulin classes. J. exp. Med. *122:* 41–58 (1965).

32 Farrands, P.A.; Pimm, M.V.; Perkins, A.C.; Hardy, J.D.; Baldwin, R.W.; Hardcastle, J.D.: Radioimmunodetection of human colorectal cancers by an anti-tumor monoclonal antibody. Lancet *ii:* 397–400 (1982).

33 Flechner, I.: The cure and concomitant immunization of mice bearing Ehrlich ascites tumors by treatment with an antibody-alkylating agent complex. Eur. J. Cancer *9:* 741–745 (1973).

34 Fodstad, O.; Olsens, S.; Pihl, A.: Inhibitory effect of abrin and ricin on the growth of transplantable murine tumor and of abrin on human cancers in nude mice. Cancer Res. *37:* 4559–4567 (1977).

35 Folkman, J.; Haudenschild, C.: Angiogenesis in vitro: implications for tumor biology; in Nossel, Vogel, Pathobiology of the endothelial cell, pp. 79–93 (Academic Press, New York 1982).

36 Froese, G.; Berczi, I.; Israels, L.G.: Tumor cell-antibody interactions. II. In vitro studies. Immunology *45:* 313–323 (1982).

37 Gaffar, S.A.; Pant, K.D.; Shochat, D.; Bennett, S.J.; Goldenberg, D.M.: Experimental studies of tumor radioimmunodetection using antibody mixtures against carcinoembryonic antigen (CEA) and colon-specific antigen-p (CSAp). Int. J. Cancer *27:* 101–105 (1981).

38 Ghose, T.; Norvell, S.T.; Guchi, A.; Cameron, D.; Bodurtha, A.; MacDonald, A.S.: Immunochemotherapy of cancer with chlorambucil-carrying antibody. Br. med. J. *iii:* 495–499 (1972).

39 Ghose, T.; Guclu, A.: Cure of a mouse lymphoma with radio-iodinated antibody. Eur. J. Cancer *10:* 787 (1974).

40 Ghose, T.; Tai, J.; Guclu, A.; Norvell, S.T.; Blair, A.H.; Aquino, J.: Use of cell

surface localizing antihepatoma antibody for tumor imaging and therapy with drug-linked antibody. Transplant. Proc. *12:* 192–194 (1980).
41 Gilliland, D.G.; Collier, R.J.; Moehring, J.M.; Moehring, T.J.: Chimeric toxins: toxic disulfide-linked conjugate of concanavalin A with fragment A from diphtheria toxin. Proc. natn. Acad. Sci. USA *75:* 5319–5323 (1978).
42 Gilliland, D.G.; Collier, R.J.: A model system involving anti-concanavalin A for antibody targeting of diphtheria toxin fragment A. Cancer Res. *40:* 3564–3569 (1980).
43 Gilliland, D.G.; Steplewski, Z.; Collier, R.J.; Mitchell, K.F.; Chang, T.H.; Koprowski, H.: Antibody-directed cytotoxic agents: use of monoclonal antibody to direct the action of toxin A chains to colorectal carcinoma cells. Proc. natn. Acad. Sci. USA *77:* 4539–4543 (1980).
44 Goldenberg, D.M.; Kim, E.E.; Deland, F.H.: Human chorionic gonadotropic radioantibodies in the radioimmunodetection of cancer and for disclosure of occult metastases. Proc. natn. Acad. Sci. USA *78:* 7754–7758 (1981).
45 Groothuis, D.R.; Fischer, J.M.; Vick, N.A.; Bigner, D.D.: Comparative permeability of different glioma models to horseradish peroxidase. Cancer Treat. Rep. *65:* 13–18 (1981).
46 Hawthorne, M.F.; Wiersema, R.J.; Takasugi, M.: Preparation of tumor-specific boron compounds. I. In vitro studies using boron-labeled antibodies and elemental boron as neutron targets. J. medicinal Chem. *15:* 449–452 (1972).
47 Hicks, J.T.; Albrecht, P.; Rapoport, S.I.: Entry of neutralizing antibody to measles into brain and cerebrospinal fluid of immunizing monkeys after osmotic opening of the blood-brain barrier. Expl Neurol. *53:* 768–779 (1976).
48 Hirschberg, H.; Rowland, G.; Thorsby, E.: In vitro immunosuppressive effects of cytotoxic agents conjugated to antihuman lymphocyte globulin. Transplantation *26:* 292–294 (1978).
49 Houston, L.L.; Nowinski, R.C.: Cell-specific cytotoxicity expressed by a conjugate of ricin and murine monoclonal antibody directed against thy 1.1 antigen. Cancer Res. *41:* 3913–3917 (1981).
50 Hurwitz, E.; Levy, R.; Mason, R.; Wilchek, M.; Arnon, R.; Sela, M.: The covalent binding of daunomycin and adriamycin to antibodies, with retention of both drug and antibody activities. Cancer Res. *35:* 1175–1181 (1975).
51 Hurwitz, E.; Maron, R.; Bernstein, A.; Wilchek, M.; Sela, M.; Arnon, R.: The effect in vivo of chemotherapeutic drug-antibody conjugates in two murine experimental tumor systems. Int. J. Cancer *21:* 747–755 (1978).
52 Johnson, J.R.; Ford, C.H.; Newman, C.E., Woodhouse, C.S.; Rowland, G.F.; Simmonds, R.G.: A vindesine-anti-CEA conjugate cytotoxic for human cancer cells in vitro. Br. J. Cancer *44:* 472–475 (1981).
53 Jung, G.; Kohnlein, W.; Luders, G.: Biological activity of the anti-tumor protein neocarzinostatin coupled to a monoclonal antibody by N-succinimidyl 3-(2-pyridyldithio)-proprionate. Biochem. biophys. Res. Commun. *101:* 599–606 (1981).
54 Kennett, H.R.; Gilbert, F.: Hybrid myelomas producing antibodies against a human neuroblastoma antigen present on fetal brain. Science *203:* 1120–1121 (1979).
55 Kim, E.E.; DeLand, F.H.; Nelson, M.O.; Bennett, S.; Simmons, E.; Alpert, E.; Goldenberg, D.M.: Radioimmunodetection of cancer with radiolabeled antibodies to α-fetoprotein. Cancer Res. *40:* 3009–3012 (1980).
56 Krolick, K.A.; Yuan, D.; Vitetta, E.S.: Specific killing of a human breast carcinoma

cell line by a monoclonal antibody coupled to the A-chain of ricin. Cancer Immunol. Immunother. *12:* 39–41 (1981).

57 Krolick, K.A.; Uhr, J.W.; Vitetta, E.S.: Selective killing of leukemia cells by antibody toxin conjugates. Implications for autologous bone marrow transplantation. Nature, Lond. *295:* 604–605 (1982).

58 Lazarus, H.; Raso, V.; Samy, T.S.A.: In vitro inhibition of human leukemic cells (CCRF-CEM) by agarose-immobilized neocarzinostatin. Cancer Res. *37:* 3731–3736 (1977).

59 Lesserman, L.D.; Machy, P.; Barbet, J.: Cell-specific drug transfer from liposomes bearing monoclonal antibodies. Nature, Lond. *293:* 226–228 (1982).

60 Levy, R.; Hurwitz, E.; Maron, R.; Arnon, R.; Sela, M.: The specific cytotoxic effects of daunomycin conjugated to antitumor antibodies. Cancer Res. *35:* 1182–1186 (1975).

61 Lin, J.Y.; Lin, L.T.; Chen, C.C.; Tserng, K.Y.; Yung, T.C.: Abrin and ricin: new antitumor substances. Nature, Lond. *227:* 292–293 (1970).

62 Locher, G.L.: Biological effects and therapeutic possibilities of neutrons. Am. J. Roentg. *36:* 1–36 (1936).

63 Mach, J.P.; Carrel, S.; Merenda, C.; Sordat, B.; Cerottini, J.C.: In vivo localization of radiolabeled antibodies to carcinoembryonic antigen in human colon carcinoma grafted in nude mice. Nature, Lond. *248:* 704–706 (1974).

64 Mach, J.P.; Forni, M.; Ritschard, J.; Buchegger, F.; Carrel, S.; Widgren, S.; Donath, A.; Alberto, P.: Use and limitations of radiolabeled anti-CEA antibodies and their fragments for photoscanning detection of human colorectal carcinomas. Oncodev. Biol. Med. *1:* 49–69 (1980).

65 Mach, J.P.; Buchegger, F.; Formi, M.; Ritschard, J.; Berche, C.; Lumbroso, J.D.; Schreyer, M.; Girardet, C.; Accolla, R.S.; Carrel, S.: Use of radiolabeled monoclonal anti-CEA antibodies for the detection of human carcinomas by external photoscanning and tomoscintigraphy. Immunol. today *2:* 239–249 (1981).

66 Mahaley, S.M.; Mahaley, J.L.; Day, E.D.: The localization of radioantibodies in human brain tumors. II. Radioautography. Cancer Res. *25:* 779–793 (1965).

67 Marrack, D.; Kubala, M.; Corey, P.; Leavens, M.; Howze, J.; Dewey, W.; Bode, W.F.; Spar, I.L.: Localization of intracranial tumors: comparative study with ^{131}I-labeled antibody to human fibrinogen and neohydrin-^{203}Hg. Cancer *20:* 751–755 (1967).

68 Mason, D.W.; Williams, A.F.: The kinetics of antibody binding to membrane antigens in solution and at the cell surface. Biochem. J. *187:* 1–20 (1980).

69 Masuho, Y.; Kishida, K.; Hara, T.: Targeting of the antiviral protein from phytolacca-americana with antibody. Biochem. biophys. Res. Commun. *105:* 462–469 (1982).

70 Mathé, G.; Loc, T.B.; Bernard, J.: Effet sur la leucémie 1210 de la souris d'une combinaison par diazotation d'améthopterine et de α-globulines de hamsters porteurs de cette leucémie par hétérogreffe. C.r. Acad. Sci (D), Paris *246:* 1626–1628 (1958).

71 McCardle, R.J.; Harper, P.V.; Spar, I.L.; Bale, W.F.; Andros, G.; Jiminez, F.: Studies with iodine-131-labeled antibody to human fibrinogen for diagnosis and therapy of tumors. J. nucl. Med. *7:* 837–847 (1966).

72 Miller, R.A.; Maloney, D.G.; Warnke, R.; Levy, R.: Treatment of B-cell lymphoma with monoclonal anti-idiotype antibody. New Engl. J. Med. *306:* 517–522 (1982).

73 Mizusawa, E.; Dahlman, H.L.; Bennett, S.J.; Goldenberg, D.M.; Hawthorne, M.F.: Neutron-capture therapy of human cancer: in vitro results on the preparation of boron-

labeled antibodies to carcinoembryonic antigen. Proc. natn. Acad. Sci. USA *79:* 3011–3014 (1982).
74 Moolten, F.L.; Capparell, N.J.; Cooperband, S.R.: Anti-tumor effects of antibody-diphtheria toxin conjugates: use of hapten-coated cells as an antigenic target. J. natn. Cancer Inst. *49:* 1057–1062 (1972).
75 Moolten, F.L.; Capparell, N.J.; Zajdel, S.H.; Cooperband, S.R.: Antitumor effects of antibody-diphtheria toxin conjugates. II. Immunotherapy with conjugates directed against tumor antigens induced by simian virus 40. J. natn. Cancer Inst. *55:* 473–477 (1975).
76 Moolten, F.L.; Capparell, N.J.; Zajdel, S.H.: Antitumor effects of antibody-diphtheria toxin conjugates. III. Cyclophosphamide-induced immune unresponsiveness to conjugates. J. natn. Cancer Inst. *55:* 709–712 (1975).
77 Moolten, F.L.; Schreiber, R.M.; Zajdel, S.H.: Antibodies conjugated to potent cytotoxins as specific antitumor agents. Immunol. Rev. *62:* 47–73 (1982).
78 Moshakis, V.; McIlhinney, R.A.J.; Raghaven, D.; Neville, A.M.: Localization of human tumour xenografts after i.v. administration of radiolabeled monoclonal antibodies. Br. J. Cancer *44:* 91–99 (1981).
79 Neuwelt, E.A.; Frenkel, E.P.; Diehl, J.T.: Clinical and animal studies of reversible osmotic blood brain barrier disruption as a new means of increasing chemotherapy delivery to central nervous system. Trans. Am. neurol. Ass. *104:* 1–5 (1980).
80 Neville, D.M.; Youle, R.J.: Monoclonal antibody-ricin or ricin A chain hybrids: kinetic analysis of cell killing for tumor therapy. Immunol. Rev. *62:* 75–91 (1982).
81 Nyström, S.: Pathological changes in blood vessels of human glioblastoma multiforme. Acta pathol. microbiol. scand. *49:* suppl. 137, pp. 9–75 (1960).
82 Oeltman, T.N.; Heath, E.C.: A hybrid protein containing the toxic subunit of ricin and the cell-specific subunit of human chorionic gonadotropin. J. biol. Chem. *254.* 1028–1032 (1979).
83 Olsnes, S.; Pihl, A.: Abrin, ricin and their associated agglutinins; in Cuatrecasas, Receptors and recognition. Series B. The specificity and action of animal, bacterial and plant toxins, p. 129 (Chapman & Hall, London 1976).
84 Pappenheimer, A.M.: Diphtheric toxin. A. Rev. Biochem. *46:* 67–94 (1977).
85 Pimm, M.V.; Jones, J.A.; Price, M.R.; Middle, J.G.; Embleton, M.J.; Baldwin, R.W.: Tumor localization of monoclonal antibody against a rat mammary carcinoma and suppression of tumor-growth with adriamycin-antibody conjugates. Cancer Immunol. Immunother. *12:* 125–134 (1982).
86 Pressman, D.; Day, E.D.; Blau, M.: The use of paired labeling in the determination of tumor-localizing antibodies. Cancer Res. *17:* 845–850 (1957).
87 Raso, V.; Griffin, T.: Hybrid antibodies as immunopharmacologic agents. Fed. Proc. *37:* 1350 (1978).
88 Raso, V.; Griffin, T.: Specific cytotoxicity of a human immunoglobulin-directed Fab'-ricin A chain conjugate. J. Immunol. *125:* 2610–2616 (1980).
89 Raso, A.; Ritz, J.; Basala, M.; Schlossman, S.F.: Monoclonal antibody-ricin A chain conjugates selectively cytotoxic for cells bearing the common acute lymphoblastic leukemia antigen. Cancer Res. *42:* 457–464 (1982).
90 Ritz, J.; Pesando, J.M.; Sallan, S.E.; Clavell, L.A.; Notis-McConarty, J.; Rosenthal, P.; Schlossman, S.F.: Serotherapy of acute lymphoblastic leukemia with monoclonal antibody. Blood *58:* 141–152 (1981).

91 Ritz, J.; Schlossman, S.F.: Utilization of monoclonal antibodies in the treatment of leukemia and lymphoma. Blood *59:* 1–11 (1982).
92 Ross, W.C.; Thorpe, P.E.; Cumber, A.J.; Edwards, D.C.; Hinson, C.A.; Davies, A.J.: Increased toxicity of diphtheria toxin for human lymphoblastoid cells following covalent linkage to anti-(human lymphocyte) globulin or its F(ab')2 fragment. Eur. J. Biochem. *104:* 381–390 (1980).
93 Sawada, H.; Tatsumi, K.; Masataka, S.; Shirakawa, S.; Makumuka, T.; Gyoichi, W.: Effects of neocarzinostatin on DNA synthesis in L1210 cells. Cancer Res. *34:* 3341–3346 (1974).
94 Schnegg, J.F.; Diserens, A.C.; Carrel, S.; Acolla, R.S.; Tribolet, N., de: Human glioma-associated antigens detected by monoclonal antibodies. Cancer Res. *41:* 1209–1213 (1981).
95 Sears, H.F.; Mattis, J.; Herlyn, D.; Häyry, P.; Atkinson, B.; Ernst, C.; Steplewski, Z.; Koprowski, H.: Phase-1 clinical trial of monoclonal antibody in treatment of gastrointestinal tumours. Lancet *i:* 762–765 (1982).
96 Seeger, R.C.: Immunodiagnosis and immunotherapy of neuroblastomas with monoclonal antibodies. J. Neuroimmunol. (in press, 1982).
97 Seeger, R.C.; Danon, Y.L.; Rayner, S.A.; Hoover, F.: Definition of a thy-1 determinant on human neuroblastoma, glioma, sarcoma, and teratoma cells with a monoclonal antibody. J. Immun. *128:* 983–989 (1982).
98 Shearer, W.T.; Turnbaugh, T.R.; Coleman, W.E.; Aach, R.D.; Philpott, G.W.; Parker, C.W.: Cytotoxicity with antibody-glucose oxidase conjugates specific for a human colonic cancer and carcinoembryonic antigen. Int. J. Cancer *14:* 539–547 (1974).
99 Sikora, K.; Alderson, T.; Phillips, J.; Watson, J.V.: Human hybridomas from malignant gliomas. Lancet *i:* 11–14 (1982).
100 Smith, T.W.; Lloyd, B.L.; Spicer, N.; Haber, E.: Immunogenicity and kinetics of distribution and elimination of sheep digoxin-specific IgG and Fab fragments in the rabbit and baboon. Clin. exp. Immunol. *36:* 384–396 (1979).
101 Steplewski, Z.; Chang, T.H.; Herlyn, M.; Koprowski, H.: Release of monoclonal antibody defined antigens by human colorectal carcinoma and melanoma cells. Cancer Res. *41:* 2723–2727 (1981).
102 Stirpe, F.; Olsnes, S.; Pihl, A.: Gelonin, a new inhibitor of protein synthesis, nontoxic to intact cells, isolation, characterization and preparation of complexes with concanavalin A. J. biol. Chem. *255:* 6947–6953 (1980).
103 Stuhlmiller, G.M.; Sullivan, D.C.; Vervaert, C.E.; Croker, B.P.; Harris, C.C.; Seigler, H.F.: In vivo tumor localization using tumor-specific monkey xenoantibody, alloantibody, and murine monoclonal xenoantibody. Ann. Surg. *194:* 592–601 (1981).
104 Thorpe, P.E.; Brown, A.N.; Ross, W.C.; Cumber, A.J.; Detre, S.I.; Edwards, D.C.; Davies, A.J.; Stirpe, F.: Cytotoxicity acquired by conjugation of an anti-Thy 1.1 monoclonal antibody and the ribosome-inactivating protein gelonin. Eur. J. Biochem. *116:* 447–454 (1981).
105 Thorpe, P.E.; Cumber, A.J.; Williams, N.; Edwards, D.C.; Ross, W.C.; Davies, A.J.: Abrogation of the non-specific toxicity of abrin conjugated to anti-lymphocyte globulin. Clin. exp. Immunol. *43:* 195–200 (1981).
106 Thorpe, P.E.; Ross, W.C.J.: The preparation and cytotoxic properties of antibody-toxin conjugates. Immunol. Rev. *62:* 119–158 (1982).

107 Trowbridge, I. S.; Domingo, D. L.: Anti-transferrin receptor monoclonal antibody and toxin-antibody conjugates affect growth of human tumour cells. Nature, Lond. *294:* 171–173 (1981).
108 Tsukada, Y.; Bischof, W. K. D.; Hini, N.; Hirai, H.; Hurwitz, E.; Sela, M.: Effect of a conjugate of daunomycin with antibodies to rat α-fetoprotein on the growth of α-fetoprotein-producing tumor cells. Proc. natn. Acad. Sci. USA *79:* 621–625 (1982).
109 Uchida, T.; Mekada, E.; Okada, Y.: Hybrid toxin of the A chain of ricin toxin and a subunit of wistaria floribunda lectin. J. biol. Chem. *255:* 6687–6693 (1980).
110 Wikstrand, C. J.; Bigner, D. D.: Expression of human fetal brain antigens by human tumors of neuroectodermal origin as defined by monoclonal antibodies. Cancer Res. *42:* 267–274 (1982).
111 Wikstrand, C. J.; Bourdon, M. A.; Pegram, C. N.; Bigner, D. D.: Human fetal brain antigen expression common to tumors of neuroectodermal origin: gliomas, neuroblastomas, melanomas. J. Neuroimmunol. *3:* 43–62 (1982).
112 Yamaizumi, M.; Mekada, E.; Uchida, T.; Okada, Y.: One molecule of diphtheria toxin fragment A introduced into a cell can kill the cell. Cell *15:* 245–250 (1978).
113 Youle, R. J.; Murray, G. J.; Neville, D. M.: Ricin linked to monophosphopentamannose binds to fibroblast lysosomal hydrolase receptors, resulting in a cell-type-specific toxin. Proc. natn. Acad. Sci. USA *76:* 5559–5562 (1979).
114 Youle, R. J.; Neville, D. M.: Anti-thy 1.2 monoclonal antibody linked to ricin is a potent cell type-specific toxin. Proc. natn. Acad. Sci. USA *77:* 5483–5486 (1980).
115 Youle, R. J.; Neville, D. M.: Kinetics of protein-synthesis inactivation by ricin-anti-thy 1.1 monoclonal antibody hybrids. Role of the ricin-B subunit demonstrated by reconstitution. J. biol. Chem. *257:* 1598–1601 (1982).

M. A. Bourdon, PhD, Duke University Medical Center, Department of Pathology, Duke University, Durham, NC 27710 (USA)

Prog. exp. Tumor Res., vol. 28, pp. 102–117 (Karger, Basel 1984)

Potential Uses of Polyamine Biosynthesis Inhibitors in Tumor Therapy

S. M. Oredsson, L. J. Marton

Brain Tumor Research Center of the Department of Neurological Surgery and Department of Laboratory Medicine, University of California, San Francisco, Calif., USA

Introduction

The polyamines putrescine, NH_2 $(CH_2)_4$ NH_2 (Pu), spermidine, NH_2 $(CH_2)_3$ NH $(CH_2)_4$ NH_2 (Sd), and spermine, NH_2 $(CH_2)_3$ NH $(CH_2)_4$ NH $(CH_2)_3$ NH_2 (Sp), are highly basic compounds that influence growth and differentiation of cells [44]. Four enzymes regulate polyamine biosynthesis in mammalian cells. Pu is formed by direct decarboxylation of *L*-ornithine by the enzyme *L*-ornithine decarboxylase (ODC). S-adenosyl-*L*-methionine decarboxylase (AdoMet DC) catalyzes the decarboxylation of AdoMet which then serves as the aminopropyl group donor in the synthesis of the higher polyamines Sd and Sp. Sd and Sp synthase catalyze the transfer of the aminopropyl group to Pu and to Sd to form Sd and Sp, respectively.

Concentrations of the polyamines and their biosynthetic enzymes are high in actively proliferating cells both in vivo and in vitro and increase rapidly when growth or differentiation is induced in resting cells [44]. Stimulation of polyamine synthesis usually precedes increases in DNA, RNA, and protein synthesis which suggests that the polyamines might be involved in the regulation of these events [44]. The polyamines have been shown to stimulate many of the biochemical reactions involved in the synthesis of DNA, RNA, and proteins [44]. Polyamines bind tightly to negatively charged structures and molecules in the cell and to nucleic acids in particular [44].

Neoplastic cells usually contain higher intracellular concentrations of polyamines than normal cells. Polyamines from tumor cells are released into extracellular fluids. Numerous studies have shown that polyamine levels are elevated in urine, sera, bone marrow aspirates, cerebrospinal fluid, and erythrocytes of many patients bearing tumors [for references see 48]. However, true clinical utility for polyamine measurement in physiological fluids (cerebrospinal fluid) has thus far been demonstrated only for medulloblastoma [38].

A number of compounds that interfere with polyamine biosynthesis, primarily by interfering with ODC and AdoMet DC have been synthesized. Methylglyoxal-bis (guanylhydrazone) (MGBG) is one of the most extensively studied inhibitors of polyamine synthesis. It is a competitive inhibitor of AdoMet DC and thus effectively blocks the formation of Sd and Sp [for references see 20, 52]. MGBG is a potent cytocidal agent, the cytotoxicity of which may not be related only to its properties as an inhibitor of polyamine biosynthesis [for references see 20, 52]. A number of inhibitors of ODC with different modes of action have been synthesized and evaluated for inhibitory activity; two of these are α-methylornithine (α-MO), a competitive inhibitor of ODC [1], and α-difluoromethylornithine (DFMO), an enzyme-activated, irreversible inhibitor of ODC [40]. These agents have shown only limited potential as cytocidal chemotherapeutic agents. However, intracellular levels of Pu and Sd are decreased by treatment with α-MO and DFMO, the result of which is a cessation or diminution of growth [36].

Polyamines in 9L Rat Brain Tumor Cells

Relationship to Growth

Cells that continuously traverse the cell cycle and cells that have been stimulated to grow have much higher activities of the enzymes in the polyamine biosynthetic pathway and have much higher polyamine levels than do noncycling cells [44]. Polyamine metabolism in asynchronously growing 9L rat brain tumor cells has been studied [22]. The activities of ODC and AdoMet DC are at their highest levels during the period of most rapid growth. A high positive correlation between the activities of these enzymes and the specific growth rate was found during the entire growth period. Sd levels fluctuated proportionally to the levels of enzyme activities.

Relationship to the Cell Cycle

Cells can be synchronized so that they may be followed through different stages of the cell cycle with regard to a variety of cellular events. Centrifugal elutriation is a technique used to isolate cells in different phases of the cell cycle based on size-density differences [26]. Cells that have been isolated can either be analyzed directly or may be grown in culture in synchrony; cells selected in one phase of the cycle, usually G_1, tend to progress through the cycle in synchrony when grown in culture after elutriation. Both approaches were used to study the relationship between polyamine levels and cell cycle phase in 9L cells [unpublished results]. The results obtained with both techniques compared favorably. Polyamine synthesis was initiated early in the cell cycle, which suggests that polyamines may be involved in the regulation of cell division, and increased progressively as cells traversed the cell cycle from G_1 to G_2/M, which suggests that polyamines may stabilize newly synthesized DNA and may be involved in mitotic processes.

Inhibition with α-MO

The effect of various concentrations of α-MO on the proliferation and intracellular polyamine content of 9L cells grown in culture was studied [59]. Treatment of 9L cells with 10–50 m*M* α-MO produced a dose-dependent inhibition of cell proliferation. During the first 24 h of incubation, α-MO-treated and control cells grew at the same rate, after which the growth rate of treated cells declined. This is in agreement with other reports which show that inhibition of proliferation does not occur until after at least one doubling time [34], the time at which significant depletion of Sd occurs, despite much earlier depletion of Pu. Pu could not be detected in 9L cells after 24 h of treatment, and Sd levels were significantly reduced after 48 h of treatment. Treatment of 9L cells with 1 m*M* α-MO did not affect cellular growth, although Pu was significantly decreased. Addition of 1 m*M* Pu to cells treated with 10 m*M* α-MO for 24 h replenished intracellular polyamine levels, and growth inhibition was prevented. The viability of 9L cells after treatment with α-MO was studied using a colony-forming efficiency (CFE) assay [11]. The CFE of treated cells did not decline significantly relative to that of untreated controls after treatment with concentrations as high as 100 m*M* of α-MO. Flow cytometric studies showed that α-MO treatment did not affect the cell cycle progression of 9L cells in any particular phase of the cycle. The doses of α-MO required for significant inhibition of 9L cell growth were quite high, and nonspecific effects could not be ruled out, even though exogenous Pu prevented the effects. For instance, 50 m*M* ornithine produced significant growth inhibition. The ornithine effect was not prevented by addition of exogenous Pu.

Inhibition with α-DFMO

Treatment of 9L cells with 1–25 m*M* of the irreversible inhibitor DFMO, effectively blocked the synthesis of Pu and Sd [58]. Compared to controls, the Sp level was either slightly elevated or unchanged. For the first 24 h treated cells grew at the same rate as controls, but thereafter significant growth inhibition occurred. Experiments in which exogenous Pu was added to culture medium after DFMO treatment suggested that the growth inhibitory effect of DFMO probably was a specific result of polyamine depletion rather than a nonspecific effect of the drug. Experiments designed to test a possible permissive role for a minimum basal level of polyamines in initiation of DNA synthesis showed that cells pretreated with DFMO and then replated in its absence do not resume proliferation until they replenish their supply of polyamines. Flow cytometric studies showed that the fraction of cells in G_1 increased until polyamine accumulation resumed, implying the presence of a G_1/S boundary inhibition. These results support the hypothesis that entry of 9L cells into S phase depends on an adequate intracellular pool of polyamines.

Inhibition with MGBG

As a means of studying whether Sd and Sp are essential for continued cell growth, the effect of the Sd and Sp synthesis inhibitor MGBG on proliferation of 9L cells in vitro was investigated [21]. Drug treatment resulted in a decrease in cellular levels of Sd and Sp, while Pu content increased. 40 μ*M* MGBG inhibited cell proliferation within one doubling time, but flow cytometric data showed that there was an almost immediate block of cell cycle traverse; cells accumulated in G_1 phase, which suggested the presence of an at least partial block in G_1. Cells that had passed G_1 completed the cycle and were blocked in G_1 during the subsequent cycle. The G_1 block suggests that DNA replication may require polyamines, in particular Sd and Sp. Addition of Sd or fresh medium were ineffective in reversing the MGBG effect.

Effects of Polyamine Depletion on Cancer Therapeutic Modalities

General Concepts

It is obvious that, because of their growth inhibitory effects, DFMO and MGBG should be investigated as cancer chemotherapeutic agents. Less obvious is the potential use of polyamine biosynthesis inhibitors as response modifiers'. Many cancer therapeutic agents are cytotoxic because they damage DNA. The polyamines are well known for their ability to stabilize DNA [5, 8, 67], and both specific [32] and nonspecific [7] interactions have been postulated. Using viscoelastometry, we have recently shown that DNA from 9L cells depleted of polyamines with DFMO is changed in conformation and is considerably more susceptible to shear than DNA from untreated cells [*Hung* et al., unpublished results]. If such destabilization of intracellular DNA could be achieved in vivo, it might be possible to alter the effects of cytotoxic agents against DNA.

X-Irradiation

Our initial investigation showed that polyamine depletion in 9L cells in monolayer culture using α-MO, DFMO, MGBG, or DFMO in combination with MGBG did not alter the dose-response curve for X-irradiation, as measured by CFE [57]. Although it seemed that our original hypothesis was incorrect, it is possible that polyamine depletion altered DNA structure, but in a manner that did not increase accessibility of the base pairs to the small hydroxyl radicals formed by X-irradiation.

1,3-bis(2-Chloroethyl)-1-Nitrosourea (BCNU)

BCNU is a chloroethylnitrosourea (CENU) that is thought to kill cells by alkylation and cross-linking of DNA [16]. The cytotoxicity of BCNU was increased in 9L cells pretreated with DFMO for 48 h before BCNU treatment [23]. At the dose used, DFMO had no cytotoxic effect on 9L cells, thus true potentiation was obtained. When Pu was added to the culture medium before BCNU treatment, the potentiation of BCNU cytotoxicity was prevented. Thus, potentiation appeared to be related to the polyamine depletion.

Cis-Platinum

Cis-Diaminodichloroplatinum (II) reacts with DNA to form Pt-base adducts on a single strand or to form base-Pt-base inter- and intrastrand cross-links [3, 53], the latter of which are thought to be cytotoxic [18]. Because the cytotoxicity of BCNU is thought to be a function of cross-link formation, we assumed that DFMO pretreatment would also potentiate cis-platinum cytotoxicity. However, we found that pretreatment of monolayer 9L cells with DFMO decreased the cytotoxicity of cis-platinum as measured by CFE [45]. The decrease in cytotoxicity could be prevented by addition of Pu after DFMO pretreatment, but before cis-platinum treatment. In cis-platinum, the two chlorine-leaving groups are separated by 3.3 Å [41], a distance very close to the interplanar distance (3.4 Å) separating DNA bases [29], and to the distance separating a number of nucleophilic groups in DNA [54]. An alteration in DNA structure secondary to polyamine depletion might change the distance between and the spatial configuration of nucleophilic groups which would make the cross-linking reaction with cis-platinum less favorable. With regard to its effects on cytotoxicity, the result of DNA destabilization secondary to polyamine depletion may be agent dependent and may confer a potential therapeutic advantage with certain agents and a potential therapeutic disadvantage with others.

BCNU and Cis-Platinum (Sister Chromatid Exchange)

The sister chromatid exchange (SCE) assay is a sensitive method for measuring damage to chromosomes [62]. SCEs are induced by compounds that damage DNA [71]. We used this assay to determine the effects of DFMO-induced polyamine depletion on the induced damage to chromosomes caused by BCNU and cis-platinum in 9L monolayer cells [66].

DFMO alone did not induce SCEs above control levels, while BCNU alone increased SCEs compared to control levels. However, in cells pretreated with DFMO and then with BCNU, the number of SCEs increased approximately twofold over the number of SCEs in cells treated with BCNU alone. Addition of Pu to DFMO-pretreated cells before BCNU treatment prevented the increase in SCEs caused by DFMO.

Cis-Platinum induces large numbers of SCEs in 9L cells. However, in 9L cells treated with DFMO and then cis-platinum, SCEs were reduced compared to treatment with cis-platinum alone; this effect was also prevented by addition of Pu. These data show that DFMO-mediated polyamine depletion modifies not only the cytotoxicity of BCNU and cis-platinum, but also modifies their induction of damage to chromosomes, and supports the postulation that these phenomena are the result of the destabilization of DNA caused by polyamine depletion.

BCNU-Cell Cycle Age Response

Although it is classified as a cell cycle nonspecific agent, BCNU is more toxic to cells in G_1 and G_2/M phases than to cells in S phase [6, 13]. Were DFMO to potentiate the cytotoxicity of BCNU predominantly in S phase, this combination would be of little value against slow-growing tumors in which few cells traverse S phase at any given time. However, we found that DFMO pretreatment of 9L cells in vitro potentiates the cytotoxic effect of BCNU in all phases of the cell cycle, particularly in G_1 and G_2/M phases [6]. Addition of Pu prevented potentiation throughout the cell cycle.

BCNU Resistance

Drug resistance is a major problem in the treatment of human tumors [19]. To investigate the possibility that DFMO might increase cell kill in resistant cells, we used a BCNU-resistant cell line [50]. DFMO pretreatment of resistant 9L cells grown in monolayer culture does not potentiate BCNU cytotoxicity. Because DFMO depletes polyamine levels by approximately the same amount in resistant and sensitive cell lines, the lack of potentiation in resistant cells cannot be attributed to a differential effect of DFMO on polyamine biosynthesis.

BCNU – Human Cell Cultures

Both BCNU-sensitive and resistant human glioblastoma cultures were pretreated with DFMO and then treated with BCNU. DFMO pretreatment potentiated BCNU cytotoxicity in BCNU-sensitive cells, but not in BCNU-resistant cells [unpublished results]. These results are very similar to results obtained in 9L cells.

Alkylation versus Carbamoylation

Nitrosoureas decompose under physiological conditions into various reactive species [42, 43, 69]; isocyanates and carbonium ions are the two major species produced. Intracellularly, isocyanates react primarily to form carbamoylates with proteins [10, 70]; the role of carbamoylation in nitrosourea-induced cytotoxicity is poorly defined [56]. Carbonium ions alkylate nucleophilic sites on both nucleic acids and proteins [10, 69]. The cytotoxicity of CENUs such as BCNU is generally thought to be the result of the initial monoalkylation of DNA and subsequent formation of DNA-interstrand and DNA-protein cross-links [14, 17, 27].

To clarify the importance of alkylation (mono- and bifunctional) and carbamoylation in the DFMO-mediated increase in BCNU cytotoxicity, we compared the effects of DFMO pretreatment on four nitrosoureas that produce different reactive decomposition products [46, 47]. 2-[3-(2-chloroethyl)-3-nitrosoureido]-*D*-glucopyranose (chlorozotocin) alkylates and cross-links [25], but does not carbamoylate [69]. 1,3-bis(trans-4-hydroxycyclohexyl)-1-nitrosourea (BHCNU) carbamoylates, but does not alkylate [69]. N-ethyl-N-nitrosourea (ENU) and N-methyl-N-nitrosourea (MNU) ethylate and methylate, respectively, nucleophilic oxygen and nitrogen in DNA, but do so with different site specificity [30, 63]; these agents do not cross-link. Polyamine depletion in 9L monolayer cultures with DFMO potentiated the cytotoxicity of chlorozotocin, but not of the other drugs [46, 47]; the cytotoxicity of BHCNU was reduced compared to treatment with BHCNU alone, and the cytotoxicities of ENU and MNU were unaffected by DFMO pretreatment. It appears that polyamine depletion does not affect monoalkylation (ENU and MNU), has an inhibitory effect on carbamoylation (BHCNU), and potentiates cross-linking (chlorozotocin).

1-(2-Chloroethyl)-3-trans-4-Methylcyclohexyl-1-Nitrosourea (MeCCNU)

This agent has a mechanism of action similar to BCNU [69]; DFMO pretreatment also potentiates cytotoxicity of MeCCNU [49]. With ^{14}C-MeCCNU [49] we showed that monoadduct formation was not increased by DFMO-pretreatment, a finding consistent with our results with ENU and MNU.

DNA Repair Enzymes

A number of enzymes are involved in the repair of alkylation damage [31]. A correlation has been found between the ability to remove alkylation damage from DNA and resistance to alkylating agents [15, 60]. An inhibition of the activity of these enzymes might be expected to increase the cytotoxicity of alkylating agents. We studied the effects of DFMO treatment on the enzymes O^6-methylguanine-DNA demethylase, 7-methylguanine-DNA glycosylase, and 3-methyladenine-DNA glycosylase in 9L cells in vitro and found no effect on the activity of these three enzymes [49].

MGBG-BCNU

DFMO causes a striking accumulation of decarboxylated AdoMet in other cell lines [35, 51] and in 9L cells [49]. Addition of Pu to the culture medium prevents accumulation of decarboxylated AdoMet. *Pegg* et al. [51] have suggested that decarboxylated AdoMet, which is thought to be a weak methylating agent, might be responsible for the potentiation of CENUs by DFMO. To investigate this possibility, we treated 9L cells with

MGBG, a polyamine inhibitor that, unlike DFMO, does not cause an increase in decarboxylated AdoMet [24]. MGBG pretreatment potentiated BCNU cytotoxicity with a dose enhancement ratio that was virtually identical to that produced by DFMO [23]. While it remains possible that the mechanism by which MGBG potentiates BCNU cytotoxicity is totally independent of the mechanism by which DFMO potentiates BCNU cytotoxicity, this appears unlikely.

Cytosine Arabinoside (Ara C)

Sunkara et al. [65] have shown that HeLa cells pretreated with DFMO accumulate in S phase and are more effectively killed by the S phase specific agent Ara C than cells treated with Ara C alone. *Rupniak and Paul* [55] and *Sunkara* et al. [64] have reported that 'normal' cell lines treated with polyamine biosynthesis inhibitors in vitro are arrested in G_1 phase and that transformed cells are arrested in S phase. This suggests the intriguing possibility that pretreatment with a polyamine biosynthesis inhibitor will protect normal tissues from the effects of S phase specific agents and will increase the relative kill of tumor cells which increases the therapeutic ratio.

DFMO pretreatment decreased the cytotoxicity of Ara C in 9L cells [unpublished results]; the effect was prevented by addition of exogenous Pu. These findings are consistent with our data that DFMO does not block 9L cells at any particular phase of the cell cycle, although there is some tendency towards accumulation in G_1 phase [58]. Because of growth inhibition, fewer 9L cells will traverse S phase at any given time. Preliminary results have shown that DFMO pretreatment of 9L cells also decreases the cytotoxicity of methotrexate, although the magnitude of the decrease is less.

The contrasting results for the same combination of agents in different cell lines points out the importance of investigating multiple models for contrasting modes of action, particularly with regard to cell cycle phenomena.

Bleomycin and Ellipticine

Bleomycins are chemotherapeutic agents that incorporate polyamines or polyamine analogues as part of their structure [68]. It is thought that this part of their structure participates in the attachment of the molecule to DNA, where further interactions are cytotoxic [9]. *Bacchi* et al. [4] reported that the ability of bleomycin to cure murine trypanosomal infections is greatly enhanced by small doses of DFMO. We found that DFMO pretreatment had no effect on the cytotoxicity of bleomycin against 9L cells.

The action of compounds that intercalate DNA might be affected by polyamine depletion. 5,11-Dimethyl-6H-pyrido[4,3-b]carbazole (ellipticine) is an intercalating agent [28] presently undergoing clinical investigation. DFMO pretreatment of 9L monolayer cultures did not affect ellipticine's cytotoxicity.

Approach to the Use of Polyamine Biosynthesis Inhibitors for Cancer Therapy

General Concepts

We have previously outlined a general approach to brain tumor chemotherapy [37] and will review here only the potential uses for polyamine biosynthesis inhibitors. More potent, less toxic, and tumor-specific agents are always actively sought. Although MGBG has potential as a cytocidal agent, DFMO is relatively nontoxic to either normal or tumor tissues as a single agent, but might be used to prevent or at least to retard tumor regrowth between the courses of cytocidal chemotherapy. An agent to be used for this purpose need not be cytocidal, but should be easily administered, minimally toxic to normal tissues, and inhibit cell proliferation.

The potentiating effect of polyamine depletion on the cytotoxicity of other chemotherapeutic agents could also be of real advantage in tumor therapy. If the cell kill obtained from a given dose of drug can be increased, or if the cell kill obtained from a dose of drug that was reduced because of toxicity to normal tissue can be maintained at the original level by using a potentiating agent, significant therapeutic advantage can be gained.

It has been shown that different polyamine biosynthesis inhibitors used in combination differentially affect cells. DFMO pretreatment causes an increased uptake of MGBG in both experimental [2] and human [61] tumors. This phenomenon is the result of the competition for cellular uptake between MGBG and the polyamines, the shift in flux secondary to decreased intracellular polyamine levels favors MGBG uptake [2]. This finding has been applied to human leukemias [61] with intriguing results and is now undergoing evaluation for human brain tumors.

Spheroids

It is virtually impossible to study the effects of combined DFMO/CENU treatment in 9L monolayer cultures for more than a single course, and it is quite difficult in the brain tumor animal models available in our laboratory. However, 9L cells grow readily as spheroids [12], a model that can be used for long-term studies of drug effects. BCNU cytotoxicity is potentiated by DFMO pretreatment in 9L spheroids; when two courses of DFMO-BCNU are administered, retardation of growth is dramatic [*Sano* et al., unpublished results].

Animal Models

Using the murine glioma 26 and the 9L rat intracerebral tumors, we studied the effects of DFMO on BCNU cytotoxicity and the delay of tumor repopulation in vivo [39]. Administration of DFMO in the drinking water for 4 days before a single intraperitoneal dose of BCNU resulted in a significant increase in survival compared with BCNU alone, showing potentiation of BCNU cytotoxicity. Administration of DFMO following a single dose of BCNU also significantly increased survival, compared with BCNU alone, showing an effect on tumor repopulation. However, these effects were not additive.

Possible Potentiation of Normal Tissue Toxicity

It is critically important that 'response modifiers' be given on the correct schedule and at the correct doses. To assure effective treatment, it is essential to optimize tumor cell kill and minimize toxicity; an inappropriate combination of agents or a poorly timed administration sequence may be detrimental rather than beneficial. For instance, normal adult rats treated with Ara C exhibit severe gastrointestinal toxicity; treatment with DFMO after administration of Ara C delays the repopulation of the gastrointestinal mucosa [33]. Similar experiments show that bone marrow depression caused by Ara C is prolonged when DFMO is administered after Ara C [*Luk and Baylin,* personal communication]. Obviously, even though DFMO administered before Ara C may be potentially beneficial, its administration immediately after Ara C may be deleterious.

Conclusions

Continued research is needed to define the combinations and sequence of administration of polyamine biosynthesis inhibitors that are most beneficial and least toxic. The potential advantages of these agents are great, and with DFMO and MGBG in clinical trial in Europe and the USA in vitro and in vivo data are particularly important. Their mechanisms of action must be understood before they can be used to full advantage. These studies will also increase our knowledge of the functions of the polyamines and of the mechanisms of action of a variety of chemotherapeutic agents.

Acknowledgments

This work was supported by NIH Program Project Grant CA-13525, American Cancer Society Grant RD–137, and travel grants (to *S.M.O.*) from the Swedish Natural Science Research Council (R-RA4685–100) and the Swedish Medical Research Council (B81-04R-6065-504106065). We thank *Neil Buckley* for expert editorial assistance.

References

1 Abdel-Monem, M.M.; Newton, N.E.; Weeks, C.E.: Inhibitors of polyamine biosynthesis. I. α-Methyl-(±)-ornithine, an inhibitor of ornithine decarboxylase. J. med. Chem. *17:* 447–451 (1974).

2 Alhonen-Hongisto, L.; Seppänen, P.; Jänne, J.: Intracellular putrescine and spermidine deprivation induces increased uptake of the natural polyamines and methylglyoxal bis(guanylhydrazone). Biochem. J. *192:* 941–945 (1980).

3 Alvarez, M.V.; Cobreros, G.; Heras, A.; Lopes Zumel, C.: Studies on cis-dichlorodiamine platinum (II) as a radiosensitizer. Br. J. Cancer *37:* 68–72 (1978).

4 Bacchi, C.J.; Nathan, H.C.; Hutner, S.H.; McCann, P.P.; Sjoerdsma, A.: Novel combination chemotherapy of experimental trypanosomiasis by using bleomycin and *DL*-α-difluoromethylornithine; reversal by polyamines. Biochem. Pharmac. *31:* 2833–2836 (1982).

5 Bachrach, U.; Eilon, G.: The effect of spermine and oxidized spermine on enzymatic degradation of DNA. Biochim. Biophys. Acta *179:* 494–496 (1969).

6 Bjerkvig, R.; Oredsson, S.M.; Marton, L.J.; Linden, M.; Deen, D.F.: The cell cycle age response of 9L cells to 1,3-bis(2-chloroethyl)-1-nitrosourea and modification by α-difluoromethylornithine. Cancer Res. *43:* 1497–1500 (1983).

7 Bloomfield, V.A.; Wilson, R.W.: Interaction of polyamines with polynucleotides; in Morris, Marton, Polyamines in biology and medicine, pp. 183–206 (Dekker, New York 1981).

8 Brown, P.E.: Some effects of binding agents on the X-irradiation of DNA. Radiat. Res. *34:* 24–35 (1968).

9 Burger, R.M.; Peisach, J.; Horwitz, S.B.: Mechanism of bleomycin action: in vitro studies. Life Sci. *28:* 715–727 (1981).

10 Cheng, C.J.; Fujimura, S.; Grunberger, D.; Weinstein, I.B.: Interaction of 1-(2-

chloroethyl)-3-cyclohexyl-1-nitrosourea (NSC 79037) with nucleic acids and proteins in vivo and in vitro. Cancer Res. *32:* 22–27 (1972).

11 Deen, D.F.; Bartle, P.M.; Williams, M.E.: Response of cultured 9L cells to spirohydantoin mustard and X-rays. Int. J. Radiat. Oncol. Biol. Phys. *5:* 1663–1667 (1979).

12 Deen, D.F.; Hoshino, T.; Williams, M.E.; Muraoka, I.; Knebel, K.D.; Barker, M.: Development of a 9L rat brain tumor cell multicellular spheroid system and its response to 1,3-bis(2-chloroethyl)-1-nitrosourea and radiation. J. natn. Cancer Inst. *64:* 1373–1382 (1980).

13 Drewinko, B.; Barlogie, B.: Age-dependent survival and cell cycle progression of cultured cells exposed to chemotherapeutic drugs. Cancer Treat. Rep. *60:* 1707–1717 (1976).

14 Erickson, L.C.; Bradley, M.O.; Ducore, J.M.; Ewig, R.A.G.; Kohn, K.W.: DNA cross-linking and cytotoxicity in normal and transformed human cells treated with antitumor nitrosourea. Proc. natn. Acad. Sci. USA *77:* 467–471 (1980).

15 Erickson, L.C.; Laurent, G.; Sharkey, N.A.; Kohn, K.W.: DNA cross-linking and monoadduct repair in nitrosourea-treated human tumour cells. Nature, Lond. *288:* 727–729 (1980).

16 Ewig, R.A.G.; Kohn, K.W.: DNA damage and repair in mouse leukemia L1210 cells treated with nitrogen mustard, 1,3-bis(2-chloroethyl)-1-nitrosourea, and other nitrosoureas. Cancer Res. *37:* 2114–2122 (1977).

17 Ewig, R.A.G.; Kohn, K.W.: DNA-protein cross-linking and DNA interstrand cross-linking by haloethylnitrosoureas in L1210 cells. Cancer Res. *38:* 3197–3203 (1978).

18 Harder, H.C.; Rosenberg, B.: Inhibitory effects of anti-tumor platinum compounds on DNA, RNA and protein synthesis in mammalian cells in vitro. Int. J. Cancer *6:* 207–216 (1970).

19 Harrarp, K.R.; Jackson, R.C.: Biochemical mechanism of resistance to antimetabolites. Antibiotics Chemother. *23:* 228–237 (1978).

20 Heby, O.; Jänne, J.: Polyamine antimetabolites: biochemistry, specificity, and biological effects of inhibitors of polyamine synthesis; in Morris, Marton, Polyamines in biology and medicine, pp. 243–310 (Dekker, New York 1981).

21 Heby, O.; Marton, L.J.; Wilson, C.B.; Gray, J.W.: Effect of methylglyoxal-bis-(guanylhydrazone), an inhibitor of spermidine and spermine synthesis, on cell cycle traverse. Eur. J. Cancer *13:* 1009–1017 (1977).

22 Heby, O.; Marton, L.J.; Wilson, C.B.; Martinez, H.M.: Polyamine metabolism in a rat brain tumor cell line: its relationship to the growth rate. J. cell. Physiol. *86:* 511–522 (1975).

23 Hung, D.T.; Deen, D.F.; Seidenfeld, J.; Marton, L.J.: Sensitization of 9L rat brain gliosarcoma cells to 1,3-bis(2-chloroethyl)-1-nitrosourea by α-difluoromethylornithine, an ornithine decarboxylase inhibitor. Cancer Res. *41:* 2783–2785 (1981).

24 Hung, D.T.; Oredsson, S.M.; Pegg, A.E.; Deen, D.F.; Marton, L.J.: Potentiation of 1,3-bis(2-chloroethyl)-1-nitrosourea cytotoxicity in 9L rat brain tumor cells by methylglyoxal-bis(guanylhydrazone), an inhibitor of S-adenosyl-*L*-methionine decarboxylase (submitted).

25 Kann, H.E.: Comparison of biochemical and biological effects of four nitrosoureas with differing carbamoylating activities. Cancer Res. *38:* 2363–2366 (1978).

26 Keng, P.C.; Li, C.K.N.; Wheeler, K.T.: Synchronization of 9L rat brain tumor cells by centrifugal elutriation. Cell Biophys. *2:* 193–206 (1980).

27 Kohn, K.W.: Interstrand cross-linking of DNA by 1,3-bis-(2-chloroethyl)-1-nitrosourea and other 1-(2-haloethyl)-1-nitrosoureas. Cancer Res. *37:* 1450–1454 (1977).

28 Kohn, K.W.; Waring, M.J.; Glaubinger, D.; Friedman, C.A.: Intercalative binding of ellipticine to DNA. Cancer Res. *35:* 71–76 (1975).

29 Langridge, R.; Wilson, L.D.; Hooper, C.W.; Wilkins, M.H.F.; Hamilton, L.D.: The molecular configuration of deoxyribonucleic acid. I. X-ray diffraction study of a crystalline form of the lithium salt. J. molec. Biol. *2:* 19–37 (1960).

30 Lawley, P.D.; Warren, W.: Removal of minor methylation products 7-methyladenine and 3-methylguanine from DNA of *Escherichia coli* treated with dimethylsulphate. Chem. Biol. Interact. *12:* 211–220 (1976).

31 Lindahl, T.: DNA repair enzymes. A. Rev. Biochem. *51:* 61–87 (1982).

32 Liquori, A.M.; Costantino, L.; Crescenzi, V.; Elia, V.; Giglio, E.; Puliti, R.; Desantis Savino, M.; Vitagliano, V.: Complexes between DNA and polyamines: a molecular model. J. molec. Biol. *24:* 113–122 (1967).

33 Luk, G.D.; Marton, L.J.; Baylin, S.B.: Ornithine decarboxylase is important in intestinal mucosal maturation and recovery from injury in rats. Science *210:* 195–198 (1980).

34 Mamont, P.S.; Bohlen, P.; McCann, P.P.; Bey, P.; Schuber, F.; Tardif, C.: α-Methylornithine, a competitive inhibitor of ornithine decarboxylase, blocks proliferation of rat hepatoma cells in culture. Proc. natn. Acad. Sci. USA *73:* 1626–1630 (1976).

35 Mamont, P.S.; Danzin, C.; Wagner, J.; Sait, M.; Joder-Ohlenbusch, A.-M.; Claverie, N.: Accumulation of decarboxylated S-adenosyl-*L*-methionine in mammalian cells as a consequence of the inhibition of putrescine biosynthesis. Eur. J. Biochem. *123:* 499–520 (1982).

36 Mamont, P.S.; Duchesne, M.; Joder-Ohlenbusch, A.; Grove, J.: Effects of ornithine decarboxylase inhibitors on cultured cells; in Seiler, Jung, Koch-Weser, Enzyme-activated irreversible inhibitors, pp. 43–53 (Elsevier, Amsterdam 1978).

37 Marton, L.J.: An approach to the use of polyamine biosynthesis inhibitors for cancer therapy. Med. Biol. *59:* 458–461 (1981).

38 Marton, L.J.; Edwards, M.S.; Levin, V.A.; Lubich, W.P.; Wilson, C.B.: CSF polyamines: a new and important means of monitoring patients with medulloblastoma. Cancer *47:* 757–760 (1981).

39 Marton, L.J.; Levin, V.A.; Hervatin, S.J.; Koch-Weser, J.; McCann, P.P.; Sjoerdsma, A.: Potentiation of the antitumor therapeutic effects of 1,3-bis(2-chloroethyl)-1-nitrosourea by α-difluoromethylornithine, an ornithine decarboxylase inhibitor. Cancer Res. *41:* 4426–4431 (1981).

40 Metcalf, B.W.; Bey, P.; Danzin, C.; Jung, M.J.; Casara, P.; Vevert, J.P.: Catalytic irreversible inhibition of mammalian ornithine decarboxylase (EC 4.1.1.17) by substrate and product analogs. J. Am. chem. Soc. *100:* 2551–2553 (1978).

41 Milburn, G.H.W.; Truter, M.R.: The crystal structures of cis-and trans-dichlorodiamineplatinum (II). J. chem. Soc. 1609–1616 (1966).

42 Montgomery, J.A.: Chemistry and structure activity studies of the nitrosoureas. Cancer Treat. Rep. *60:* 651–664 (1976).

43 Montgomery, J.A.; James, R.; McCaleb, G.S.; Johnston, T.P.: The modes of decomposition of 1,3-bis(2-chloroethyl)-1-nitrosourea and related compounds. J. med. Chem. *10:* 668–674 (1967).

44 Morris, D.R.; Marton, L.J.: Polyamines in biology and medicine (Dekker, New York 1981).

45 Oredsson, S.M.; Deen, D.F.; Marton, L.J.: Decreased cytotoxicity of cis-diamminedichloroplatinum (II) by α-difluoromethylornithine depletion of polyamines in 9L rat brain tumor cells in vitro. Cancer Res. *42:* 1296–1299 (1982).

46 Oredsson, S.M.; Deen, D.F.; Marton, L.J.: The influence of polyamine depletion on alkylation and carbomoylation induced cytotoxicity; in Bachrach, Kaye, Chayen, Advances in polyamine research, vol. 4, pp. 41–43 (Raven Press, New York 1983).

47 Oredsson, S.M.; Deen, D.F.; Marton, L.J.: The influence of polyamine depletion caused by α-difluoromethylornithine, an enzyme-activated irreversible inhibitor of ornithine decarboxylase, on alkylation- and carbamoylation- induced cytotoxicity in 9L rat brain tumor cells in vitro. Cancer Res. (in press).

48 Oredsson, S.M.; Marton, L.J.: Polyamines: the elusive cancer markers. Clin. Lab. Med. *2:* 507–518 (1982).

49 Oredsson, S.M.; Pegg, A.E.; Deen, D.F.; Marton, L.J.: Possible factors in the potentiation of 1-(2-chloroethyl)-3-trans-4-methylcyclohexyl-1-nitrosourea cytotoxicity by α-difluoromethylornithine in 9L rat brain tumor cells (submitted).

50 Oredsson, S.M.; Tofilon, P.J.; Feuerstein, B.G.; Deen, D.F.; Rosenblum, M.L.; Marton, L.J.: Differential potentiation of 1,3-bis(2-chloroethyl)-1-nitrosourea by α-difluoromethylornithine in chloroethylnitrosourea-sensitive and resistant 9L rat brain tumor cells in vitro. Cancer Res. (in press).

51 Pegg, A.E.; Pösö, H.; Shuttleworth, K.; Bennett, R.A.: Effect of inhibition of polyamine synthesis on the content of decarboxylated S-adenosylmethionine. Biochem. J. *202:* 519–526 (1982).

52 Porter, C.W.; Dave, C.; Mihich, E.: Inhibition of S-adenosylmethionine decarboxylase as an approach in cancer therapeutics; in Morris, Marton, Polyamines in biology and medicine, pp. 407–436 (Dekker, New York 1981).

53 Roberts, J.J.; Pascoe, J.M.: Cross-linking of complementary strands of DNA in mammalian cells by antitumor platinum compounds. Nature, Lond. *235:* 282–284 (1972).

54 Roos, I.A.G.; Thomson, A.J.; Mansy, S.: Interaction of platinum compounds with dinucleotides. J. Am. chem. Soc. *96:* 6484–6491 (1974).

55 Rupniak, H.T.; Paul, D.: Selective killing of transformed cells by exploitation of their defective cell cycle control by polyamines. Cancer Res. *40:* 293–297 (1980).

56 Schmall, B.; Cheng, C.J.; Fujimura, S.; Gersten, N.; Grunberger, D.; Weinstein, I.B.: Modification of proteins by 1-(2-chloroethyl)-3-cyclohexyl-1-nitrosourea (NSC 79037) in vitro. Cancer Res. *33:* 1921–1924 (1973).

57 Seidenfeld, J.; Deen, D.F.; Marton, L.J.: Depletion of intracellular polyamine content does not alter the survival of 9L rat brain tumor cells after X-irradiation. Int. J. Radiat. Biol. *38:* 223–229 (1980).

58 Seidenfeld, J.; Gray, J.W.; Marton, L.J.: Depletion of 9L rat brain tumor cell polyamine content by treatment with *D,L*-α-difluoromethylornithine inhibits proliferation and the G_1 to S transition. Expl Cell Res. *131:* 209–216 (1981).

59 Seidenfeld, J.; Marton, L.J.: Effects of *DL*-α-methylornithine on proliferation and polyamine content of 9L rat brain tumor cells. Cancer Res. *40:* 1961–1966 (1980).

60 Shiloh, Y.; Becker, Y.: Kinetics of O^6-methylguanine repair in human normal and ataxia telangiectasia cell lines and correlation of repair capacity with cellular sensitivity to methylating agents. Cancer Res. *41:* 5114–5120 (1981).

61 Siimes, M.; Seppänen, P.; Alhonen-Hongisto, L.; Jänne, J.: Synergistic action of two polyamine antimetabolites leads to a rapid therapeutic response in childhood leukemia. Int. J. Cancer *28:* 567–570 (1981).

62 Solomon, E.; Bobrow, M.: Sister chromatid exchange – a sensitive assay of agents damaging human chromosomes. Mut. Res. *30:* 273–277 (1975).

63 Sun, L.; Singer, B.: The specificity of different classes of ethylating agents toward various sites in HeLa cell DNA in vitro and in vivo. Biochemistry *12:* 1795–1802 (1973).

64 Sunkara, P.S.; Fowler, S.K.; Nishioka, K.: Selective killing of transformed cells in combination with inhibitors of polyamine biosynthesis and S-phase specific drugs. Cell Biol. int. Rep. *5:* 991–997 (1981).

65 Sunkara, P.S.; Fowler, S.K.; Nishioka, K.; Rao, P.N.: Inhibition of polyamine biosynthesis by α-difluoromethylornithine potentiates the cytotoxic effects of arabinosyl cytosine in HeLa cells. Biochem. biophys. Res. Commun. *95:* 423–430 (1980).

66 Tofilon, P.J.; Oredsson, S.M.; Deen, D.F.; Marton, L.J.: Polyamine depletion influences drug-induced chromosomal damage. Science *217:* 1044–1046 (1982).

67 Tsuboi, M.: On the melting temperature of nucleic acids in solution. Bull. chem. Soc. Japan *37:* 1514–1522 (1964).

68 Umezawa, H.: Chemistry and mechanism of action of bleomycin. Fed. Proc. *33:* 2296–2302 (1974).

69 Wheeler, G.P.: A review of studies on the mechanism of action of nitrosoureas. Am. chem. Soc. Symp. Ser. *30:* 87–119 (1976).

70 Wheeler, G.P.; Bowdon, B.J.; Shuck, R.F.: Carbamoylation of amino acids, peptides, and proteins by nitrosoureas. Cancer Res. *35:* 2974–2984 (1975).

71 Wolff, S.: Sister chromatid exchange. A. Rev. Genet. *11:* 183–201 (1977).

L.J. Marton, MD, Brain Tumor Research Center, Department of Neurological Surgery, University of California, San Francisco, CA 94143 (USA)

Prog. exp. Tumor Res., vol. 28, pp. 118–135 (Karger, Basel 1984)

Immunotherapy of Patients with Glioma: Fact, Fancy, and Future

M. S. Mahaley, Jr., G. Yancey Gillespie

Division of Neurological Surgery, University of North Carolina, Chapel Hill, N.C., USA

Considering the resurgence of emphasis on natural foods, natural childbirth, natural shampoos, and the like, the potential use of 'natural' defenses to fight cancer is an attractive concept. The possibility of mobilizing such defenses has fostered several areas of active investigation. Among these approaches could be included immunotherapy.

The most efficacious treatment available now for patients with anaplastic gliomas consists of radical surgical resection of the tumor followed by concurrent irradiation and systemic chemotherapy [58]. Other than the obvious difficulties involved in treating a relatively inaccessible neoplasm growing in a rigidly confined space, there are several strategic problems of a temporal nature inherent in brain tumor therapy as it is currently practiced. The ratio of actual treatment time to duration of the disease is quite low. Surgical resection cannot be performed repeatedly throughout the course of the disease; furthermore, it is probably the form of treatment with the most finite temporal limitations. Radiation therapy, as it is now most commonly practiced, exposes all of the cells within the brain to radiation for a brief interval each day over a period of 6–7 weeks. Even with innovative hyperfractionation schemes, the cumulative dose stresses the tolerance of normal brain cells, making them potentially vulnerable to the harmful effects of radiation. Chemotherapy itself is not a modality that can be applied without interruption at the present time, and the current concepts regarding both the route and timing of chemotherapeutic exposure are mostly empirical, being determined mainly by the patient's tolerance as well as by the convenience to the physician. Other novel modalities, such as hyperthermia, radiosensitizers, interstitial irradiation, and low-dose 'continuous' radiation therapy still represent discontinuous forms of therapy, if not, indeed, one-time treatments only [4, 7, 48].

There are also tactical problems. Each patient presents a unique combination of factors that influence both the appropriateness of a given therapeutic modality and the degree of clinical response; these factors include the age, sex, general medical health, tolerance to treatment, and degree of anaplasia of the glioma. The sensitivity of the tumor cells targeted for destruction is a crucial factor – not as far as surgical effectiveness is concerned, but rather from the standpoint of efficacy of radiation therapy, chemotherapy, and perhaps other adjunctive forms of treatment. Simply reviewing the survival tables shows quite clearly that in most cases in which the therapy spares irreversible damage to eloquent brain tissues, a number of malignant cells escape destruction.

While immunotherapy may prove to be no more of a panacea than surgery, radiation therapy and chemotherapy have been, elements of the immune system provide exquisitely selective routes for potentially seeking out and destroying malignant cells. Thus, clinical immunotherapeutic evaluations afford both the opportunity and the potential for: (1) a more thorough attack on neoplastic cells that have shared antigens, (2) ongoing, continuous therapy in individuals sensitized to tumor antigens, (3) a specificity of cytotoxic attack with complete sparing of normal tissues, and (4) a variety of potential therapeutic mechanisms, such as cellular, humoral, and/or antibody-dependent cell-mediated immunity. The influence that serum 'blocking factor', helper cells, and suppressor cells may exert on immunotherapeutic measures has not been defined, but certainly warrants investigation. The efficacy of conventional therapies continues to be compromised by the heterogeneity of cells within the intracranial neoplasm [5]. From patient to patient, and even within a single tissue sample, there are commonly gross and microscopic morphological differences among tumor cells. Moreover, the varied patterns of growth among these tumors in vivo are reflected in the morphologic and functional differences that exist in the clonal derivatives produced in vitro from many of these tumors. Within a patient's tumor, there may be areas of varying degrees of mitotic activity, cells that possess certain markers (S100 protein, glial fibrillary acidic protein, fibronectin) and those that do not, and cells that may or may not respond to extrinsic stimuli, such as cyclic AMP. This heterogeneity, if manifest also as antigenic differences, could confound the immunotherapeutic approach to the control of glioma growth, as well. Within the neuropathologic entity of anaplastic glioma, or even within an individual tumor, antigen profiles may not demonstrate a common antigenic denominator. This heterogeneity could magnify the complexity of achieving immunologic recognition of the target cells and of obtaining the host response required for an effective attack on the tumor. Despite these caveats, an evaluation of immunotherapeutic measures against gliomas must take into consideration the patient's general immune status as well as any specific, spontaneously generated immune response.

Immunobiology

The immunologic profile of patients with anaplastic gliomas at the time of surgical intervention, when the diagnosis is first established, has been well documented [10, 31, 32, 46]. Subnormal immune competence is reflected in: depressed delayed hypersensitivity cutaneous reactions to recall antigens; a diminished percentage and number of both total lymphocytes and T cells in the peripheral blood, and an impaired mitogen-induced blastogenic response by mononuclear cells in the peripheral blood. However, the relative percentage of cytotoxic/suppressor T cells in the peripheral blood often appears to be increased, while that of helper T cells may be either decreased or not significantly reduced. Furthermore, the quantity of B cells and monocytes in peripheral blood is usually within normal limits.

Evidence supporting the notion that patients with intracranial gliomas develop cell-mediated immune reactivity to their tumors is conflicting and controversial. Several studies have documented that the areas of perivascular infiltration frequently seen in histologic sections of anaplastic gliomas are composed largely of T lymphocytes and macrophages – two cell types that play decisive roles both as effectors and regulators of cell-mediated immunity [40, 45]. The implication of their presence in these areas, sometimes in juxtaposition with tumor cells, depends on their functional state. We would like to believe that these two cell types could be mobilized in increasing numbers and in a more active functional state to destroy tumor cells in situ. In vitro analyses of mononuclear cells in the peripheral blood of patients with glioma, however, have so far failed to demonstrate conclusively that they are capable of an antitumor effect [61]. There are many explanations for the apparent paradox of the presence of intratumoral immune-related cells and the absence of peripheral antitumor activity, that would leave intact the notion that the tumor patient reacts with an antiglioma immune response. The most obvious observation in support of this notion is that, even in well-defined animal models, there is poor correlation between immunologic reactivity by effector cells isolated in vitro from peripheral lymphoid tissues and that by effector cells isolated directly from the tumor [60]. In the absence of functional assays on intraglioma immune-related cells, we do not know if the T lymphocytes and macrophages infiltrating some gliomas possess antitumor effector activity, if they act as suppressor cells down-regulating any immune response to the neoplasm, or if they exhibit some other functional capability.

From the standpoint of their capability for humoral immune responses, it is generally held that patients with gliomas have essentially normal levels of immunoglobulins in serum, with the exception of a slightly elevated IgM level, at the time of diagnosis. There have been only a few studies documenting specific humoral immune reactions by brain tumor patients, as assessed in vitro. Several studies purport to demonstrate binding of serum antibodies to autologous and allogeneic glioma cells in vitro [1, 16, 19, 25, 37, 44]. The patient's inability to mount and sustain an antibody response to repeated (bimonthly) vaccination with such foreign antigens as influenza or tetanus would suggest a basic impairment that possibly could extend to any putative glioma-associated antigens [31]. As with cell-mediated immunity, there is evidence of the body's attempts to mobilize an antibody response. Immunoglobulin has been detected binding to both glioma cells and Fc receptor-positive inflammatory immune cells within biopsy specimens of tumor tissue [61]. Finally, in vitro assays have revealed serum 'factors' from glioma patients that block the capacities of mononuclear cells in the peripheral blood to undergo mitogen-induced blastogenesis [12, 64] and to mediate cytotoxic effects on cultured glioma cells [12, 26, 62]. Although it has been suggested that these factors may be antiglioma antibodies or immune complexes, unequivocal direct evidence to support this theory is lacking. However, studies of circulating immune complexes in patients with malignant glioma indicated that those patients with elevated levels of immune complexes had a significantly shorter survival time [18, 36].

The presence of Fc receptor-positive immune cells presumably bearing immunoglobulin within tumors raises the possibility that an antitumor effect could be mediated by means of antibody-dependent cellular cytotoxicity. However, the few in vitro studies of antibody-dependent cellular cytotoxicity by peripheral blood mononuclear cells of glioma patients have shown infrequent occurrences of specific reactions [62].

It is not possible at this time to say if any of these observations on immune response is related to the development of intracranial glioma, is a consequence of the glioma, or is simply an unrelated phenomenon. In the individual patient, however, certain of these observations do seem to be valuable in predicting tumor recurrence before clinical signs of deterioration develop, to the extent that an immune index can be determined [11]. More importantly, the study by *Brooks* et al. [11] unequivocally demonstrated the usefulness of serially monitoring changes in the immune status of patients with malignant gliomas. Nonetheless, isolated assays performed without adequate reference points within a specified time-frame would serve only to confuse interpretation of poorly understood, complex phenomena. Documentation of serial changes in immune function during exploratory efforts at immunotherapy is needed to unravel the basis for success or failure in each individual case.

Immunotherapy

All of the classic approaches to immunologic therapy – namely, adjuvant (immunorestorative) therapy, adoptive immunotherapy, passive serotherapy, and active immunotherapy – have been attempted, singly or in combination, in patients with anaplastic gliomas. None of these trials has so far identified an immunotherapeutic approach that is effective in boosting specific or general immune reactions or in improving overall survival. Only two of these clinical trials could be considered to be randomized prospective studies, and both of these have had negative results [9, 33].

The underlying concept in adjuvant immunotherapy has been the attempt to stimulate optimally the immunologic responsiveness of the patient to the tumor, overcoming what has been perceived to be a subnormal level, or even failure, of immune response. The first generation of immunorestorative, immunoaugmenting, or immunomodulating agents used in patients with brain tumors were microbial or synthetic substances known to evoke strong immune responses. Both Bacille de Calmette-Guérin vaccine (BCG) and *Corynebacterium parvum* have been used separately in trials of immunorestorative therapy in small series of glioma patients. Unfortunately, no definable therapeutic responses were detected [17, 51, 54]. *C. parvum* given intravenously in patients with significant intracranial tumor masses precipitated some problems with increased intracranial pressure during the course of the febrile response that ensued. This observation may be pertinent to other therapies of a similar nature.

Levamisole is an excellent example of the type of agent that is purported to restore immunologic capabilities of the tumor-bearing host whose immunologic reactivity is below normal. Its positive immunorestorative effects in several animal tumor systems and in limited clinical trials have been encouraging. However, in a randomized, prospective study involving 100 patients, it was found that levamisole did not alter the survival of patients with anaplastic gliomas who received it during the course of radiation therapy and chemotherapy [33]. Moreover, levamisole immunostimulation did not boost the general immune competence or produce any in vitro evidence of specific cellular or humoral cytotoxicity.

As we expand our understanding of the mechanisms by which immune-related cells initiate and govern each other's activities, immunotherapeutic efforts are turning away from the use of microbial or synthetic substances that nonspecifically activate the immune system. Instead, fledgling efforts are under way to activate selectively those immune mechanisms that are thought to be desirable by using partially purified preparations of the soluble mediators by which immune systems components communicate.

The recent advances in basic research on lymphokines, cytokines, monoclonal antibodies and other biologic substances that exert specific effects on the immune system have given rise to a new, specialized branch of cancer investigation, namely, biologic response modifier research. A new program within the Division of Cancer Treatment of the National Cancer Institute has been instituted and, to confirm the importance attributed to engendering research in this area, a new journal has been spawned to disseminate rapidly new laboratory techniques, preclinical studies, and clinical trials related to biologic response modifiers.

Lymphokines, soluble substances produced by stimulated or activated lymphocytes, constitute an important form of intercellular communication by which the activities of immune-related cells are initiated, regulated, and suppressed. At least two lymphokines are being investigated as a second generation of immunomodulating agents. Thymosin $\alpha 1$ is a low-molecular-weight peptide that is a potent inducer of T lymphocytes. Based on studies in mice and humans, the principal effect of thymosin $\alpha 1$ would appear to be reconstitution of an immunologic deficit in T-cell function, rather than augmentation of normal T-cell function to higher levels. Thymosin $\alpha 1$ is believed to act at a very early stage of T-cell development to cause a functional maturation of the immature T cell to T lymphocytes. There is some evidence that suppressor cell activity in tumor-bearing animals and cancer patients may be decreased by thymosin therapy [52]; however, the dosage used may be an important factor. There are reports of dose-dependent activation of suppressor cells in animal tumor model systems by thymosin α1 [35]. In the few clinical trials involving patients with nonneuroglial malignancies, the ability of thymosin α1 to increase survival was seen only in those cases in which the patients initially had depressed T-cell levels. Patients with relatively normal T-cell levels did not benefit from thymosin therapy [14]. Only one study [43] of patients with malignant brain tumors has involved thymosin therapy, the final results of which have not been reported. Thymosin α1 has also been found to stimulate lymphocytes to produce other soluble mediators such as macrophage inhibitory factor and interferon [24].

Interferon is a complex family of secreted glycoproteins that may possess antiviral, immunoregulatory, and antiproliferative functions. Its mechanism of action in tumor growth inhibition may be either a direct cytotoxic effect on the neoplastic cells, an embellishment of the immune system, activation of nonspecific antitumor effector cells, or some combination of these. Interferon has been shown to inhibit the growth of glioma cells in tissue culture [2, 27, 28]. It has been administered intravenously and intratumorally in a small number of patients with intracranial tumors [41, 49].

At the University of North Carolina, we have just completed a phase I study to evaluate toxicity and patients' tolerance to progressively increased intravenous dosages of a partially purified human leukocyte α interferon. In this initial series of 9 patients, febrile responses were common during interferon administration, but no untoward neurologic sequelae or major organ toxicity has been observed when dosages of $10–30 \times 10^6$ units/m^2 were administered on each of 3 consecutive days for 3 consecutive weeks. Interferon levels in serum and in some cases in cerebrospinal fluid and/or within the tumor have been assayed in these patients. When these analyses are completed, a phase II study (20 patients with recurrent malignant glioma) will be conducted using the highest tolerable dose from the phase I study. The phase II trial will examine specifically the clinical response to therapy as determined by the neurologic examination, Karnofsky functional rating, and CT brain scan. In addition, several variables of immune status will be evaluated, including levels and in vitro functions of mononuclear cells in peripheral blood (natural killer lymphocytes, monocytes, and T-lymphocyte subsets).

Adoptive immunotherapy of tumor-bearing laboratory animals with sensitized lymphocytes has been widely shown to provide a substantial protective effect against tumor progression. Based on a clear demonstration of an apparent deficit in cell-mediated immune responsiveness in patients with brain tumors, it would seem logical to attempt to resupply whatever factor is lacking. In the past, at least four groups of investigators have injected relatively large numbers of autologous or homologous leukocytes directly into the intrathecal or intratumoral space in patients with brain tumor, in the hope that the close proximity or juxtaposition of cellular elements of the immune system would result in tumor cell destruction [42, 54, 55, 63]. Patients have tolerated the therapy reasonably well, but dramatic responses to it have been the exception. Perhaps the main reason for failure has been the lack of a sufficiently sensitized pool of lymphocytic effector cells.

It is only during the past few years that both biochemical and functional parameters have been established for some of the soluble mediators that function to generate the clones of specifically sensitized effector lymphocytes [21, 47, 53, 65]. Based on this knowledge, the technical capability is now readily available to produce in vitro an unlimited number of killer T lymphocytes from each glioma patient sensitized specifically against his own tumor [56, 57]. It has been shown recently that natural killer lymphocytes can also be grown in long term culture using medium containing these same growth factors. In theory, a small number of peripheral-blood mononuclear cells or lymphocytes isolated from tumor tissue removed at surgery [56] could be incubated in vitro with the patient's tumor cells. Under proper conditions, a portion of these cells should respond to the tumor cells by blastogenic transformation. In fact, two research groups have already documented that glioma patients' peripheral-blood mononuclear cells undergo blast transformation following cocultivation in vitro with autochthonous glioma cells [15, 19]. What has been lacking in the past is the means by which to increase greatly the numbers of these responding lymphocytes. By providing sufficient amounts of T-cell growth factor – otherwise known as IL-2 for interleukin-2 – it is theoretically, although probably not practically, possible to generate greater than 10^9 killer T cells within 30 days. These T lymphocytes express cytolytic activity in vitro and continue to proliferate as long as IL-2 is in sufficient quantity [47]. Thus, T cells are not transformed by this process. The immunotherapeutic implications are obvious: Based on this technology, the patient could serve as his own source of specifically sensitized lymphocytes for 'auto-adoptive' immunotherapy. The lymphocytes can be cloned in vitro and then expanded to separate suppressor T cells from killer T cells, thus ensuring that unwanted, and possibly deleterious, immunosuppressive T-cell subsets are deleted [57]. Moreover, the cytotoxic specificity could be easily confirmed by extensive in vitro testing to eliminate those killer T-cell clones that attack normal neuroglial elements. Since the lymphocytes grow rapidly and continuously in vitro, an initial seed population would provide a readily renewable source of antitumor lymphocytes for regularly spaced therapeutic attempts. Most of the cultured cells are in a blast phase, and it has been observed that lymphoblasts 'home' best to sites of antigen deposition or inflammation [39]. It is conceivable that systemic administration of purified preparations of human IL-2 could be used to extend the half-time of infused lymphoblasts, thereby increasing their in vivo effectiveness [13]. Phase I clinical trials involving infusion of in vitro activated autologous lymphocytes into patients with tumors in the soft tissue have been reported [38].

This potential does not suggest, however, that this approach be tried immediately in a clinical setting. Although infusion of autologous and even allogenic leukocytes into the tumor bed has been reasonably well tolerated, the addition of an enriched population of specifically sensitized killer T cells could hold possibly serious consequences of both an immediate and a long-term nature for the patient. When this immunologic technique has demonstrated efficacy with low risk in acceptable animal models of intracranial neoplastic disease – then, and only then, should its efficacy be investigated clinically. Such studies must be undertaken with an acute awareness that gliomas, in addition to their remarkable degree of heterogeneity, may successfully evade such a therapeutic approach by the process of antigenic modulation.

The first study of passive serotherapy involved the administration of xenogeneic (rabbit) antiglioma antibody globulin to a small series of patients with anaplastic gliomas [33]. Antibody localization within human glioma tissues was demonstrated by radioisotope brain scanning and by direct measurements of tissue radioactivity. However, this was a limited trial because of (a) the exhaustive absorptions required after production of the antiglioma antibody in order to obtain a partially purified product; (b) the finite, and nearly infinitesimal, quantity of antibody available for study, and (c) the unknown consequences of using heterologous antibody repeatedly. The emergence of monoclonal antibody techniques now may serve to rejuvenate interest in passive serotherapy. Although this topic is extensively discussed elsewhere in these volumes (vol. 27, pp. 132–153; vol. 28, pp. 79–101), it is pertinent that presumed target antigens are being identified in human neuroglial tumors with the use of monoclonal antibodies. A neuroectodermal antigen shared by melanoma, glioma, and neuroblastoma cells has been identified [50]. In addition, a 'glioma mesenchymal extracellular matrix' antigen has been identified in studies with another monoclonal antibody derived by immunizing mice with cultured tumor cells from human glioma tissue [59]. It is conceivable that monoclonal antibodies directed against 'common glioma-distinctive' antigens will become available in quantities sufficient not only to study localization within glioma tissue, but also to expand the prospects of serotherapy by permitting use of the antibody either for direct attack (humoral immunotherapy) on the glioma cell or as a carrier of destructive agents (radionuclides, chemotherapeutic drugs, or polyhedral boron for localized neutron beam therapy).

Active immunotherapy probably remains the most complex, yet attractive, approach to immunologic control of intracranial tumors. In 1960, *Bloom* et al. [8] placed autologous implants in the thighs of a small group of patients. *Grace* et al. [22] performed similar studies at about the same time. Tumors grew locally and metastasized to regional nodes; no therapeutic benefits derived from this approach. However, regression of local implants was observed in 4 patients, 2 of whom subsequently developed positive delayed hypersensitivity skin-test reactions to extracts of their tumors. In 1970, *Trouillas and Lapras* [55] immunized 20 patients with autologous tumor tissue mixed with Freund's complete adjuvant. Although conclusions regarding therapeutic implications cannot be drawn from this series, serum samples from 14 of the 20 patients developed identifiable immunodiffusion precipitation reactions with the autologous glioma extracts. In 1972, *Febvre* et al. [20] reported the development of positive delayed hypersensitivity cutaneous reactions in 23 of 24 glioma patients to glioma cell lines following immunotherapy with autologous tissues. A randomized prospective controlled study was performed by *Bloom* et al. [9] using autologous irradiated glioma cells. Of the 27 patients who were actively immunized, 10 received multiple inoculations. Compared to a series of control patients who were not immunized, the inoculated patients had no increase in survival time. Patients in both groups were treated with radiation therapy. No delayed hypersensitivity cutaneous responses to glioma extracts developed after immunization. It should be noted that in no case were these patients selected on the basis of their having general or specific immunocompetence before their entry into the study, and an adjuvant was not incorporated with the inoculated glioma cells. This, of course, could have influenced the study results.

Whenever active immunization with neuroglial tissue is proposed, a prime consideration must be the possibility of the development of allergic encephalomyelitis, particularly when autologous tissue is used and, most certainly, when the tissue is incorporated with an adjuvant. At least 1 case of this sort probably developed in the series of patients treated by *Trouillas and Lapras* [55]. The inoculation of nonhuman primates with human glioma tissue mixed with an adjuvant can produce allergic encephalomyelitis [6]. The major encephalitogens are certain proteolytic degradation products of myelin basic protein (MBP). Protection of experimental laboratory animals from potential encephalitogenesis can be accomplished, to some extent, by one of two methods [23]. Firstly, if the animal is immunized initially without adjuvant, a protective immune response ensues that prevents the development of an autoimmune histopathology when further antigenic challenge, even with an adjuvant, is presented. This technique has been used successfully in nonhuman primates immunized with tissue freshly obtained from a human glial tumor [6]. However, it should be noted that these studies were conducted exclusively with animals that had an intact blood-brain barrier, i.e., they involved no intracranial neoplasms or traumatic insults to the brain such as surgery might produce. The other means by which a protective effect has been produced involves the enzymatic cleavage of MBP with chymotrypsin to produce 'nonencephalitogenic fragments'. The key to success with this approach would depend on ensuring somehow that an adequate enzymatic degradation of all available MBP in a glioma tissue sample has occurred. Without this assurance, which we presently have no means to obtain, active immunization would be exceedingly risky, at best; and even if the necessary immunologic machinations were developed, the prospect of human glioma tissue possessing antigens capable of evoking an allergic response within CNS tissues is a source of concern. This problem can be avoided completely, of course, by immunizing with malignant glioma cells that have been subcultured at least once in vitro and shown to be free of MBP.

In a pilot study that is virtually completed [29], a series of 20 patients were selected, within 3 weeks after surgical resection of an intracranial anaplastic glioma, for entry into an active immunization protocol. To be eligible for this study, the patient's immune profile had to be suggestive of a reasonable, although impaired, level of immunocompetence: a peripheral blood lymphocyte count greater than 1,000/mm^3, a concentration of peripheral-blood T lymphocytes equal to at least half that of normal individuals, and delayed hypersensitivity cutaneous responses to at least one out of four recall antigens. In addition, selected patients could not be receiving steroid therapy, except during the immediate perisurgical period, and had to have Karnofsky ratings equal to or greater than 70. Each patient was immunized using one of two well-characterized, long-term glioma cell lines that had been tested extensively in nonhuman primates and found not to produce allergic encephalomyelitis when inoculated together with adjuvant. As an initial immunizing dose, these 20 patients received both 10^8 glioma cells that had been lethally irradiated with 20,000 rad and 500 μg of BCG-cell wall; the mixture was inoculated in divided aliquots into four subcutaneous sites. At the same time, each patient began to take levamisole (2.5 mg/kg) orally for 3 days biweekly. At monthly intervals thereafter, each patient was reinoculated with irradiated glioma cells alone. Each patient also received a full course of irradiation to the whole head (6,000 rad total over 6 weeks) combined with BCNU chemotherapy (80 mg/m^2/day intravenously, 3 times every 8 weeks) that were both begun 1 month after the initial immunization. Briefly, this study was designed (a) to use allogeneic, tissue-cultured glioma cells [5] that did *not* produce allergic encephalomyelitis in nonhuman primates; (b) to include the use of an adjuvant that had *not* been part of the only other prospective active immunization study [9]; (c) to prescreen patients to document some degree of immune competence, and (d) to initiate the immunotherapy at least a month before beginning conventional postoperative adjunctive treatment (radiation and chemotherapy) that could adversely influence immune reactivity.

General immune reactions, such as lymphocyte and T-cell counts in peripheral blood, delayed hypersensitivity cutaneous responses to the recall antigens, and serum IgM levels, were followed serially in these patients. There was no divergence in these parameters from what would be expected without active immunotherapy. 11 patients were inoculated with the D-54MG glioma cell line and 9 patients with the U-251MG cell line. More of the patients inoculated with the D-54MG cell line developed significant reactions locally at the site of the original inoculum than did those who received U-251MG. Because fetal bovine serum (FBS) antigens become intercalated in the membranes of the tissue cultured cells used for inoculation, almost every patient generated serum antibodies against FBS antigens, the antibody activity tending to increase with time [30]. Most of the patients also developed significant serum antibodies against histocompatibility locus antigens (HLA). After absorption to deplete serum antibodies to FBS and HLA, several patients continued to show significant serum antibody activity against the original cell line used for immunization. However, after all nongliomatous antibody activity was finally absorbed, only 1 of these patients still had significant serum antibody levels against the original cell line as well as to other, but not all, glioma cell lines [30]. The exhaustively absorbed sera from this single patient also exhibited cytotoxicity against the cell line used for immunization – and to a much greater extent than we have ever observed in unimmunized patients with intracranial gliomas [62]. The unique serological response in this patient was intriguing. Although this phase I toxicity trial was not designed to provide statistically meaningful survival data, the life table of percent survival for the group of patients inoculated with U-251MG indicated a prolongation of survival when compared with that of patients inoculated with D-54MG, or with a comparable group of patients with identical diagnoses who were treated with levamisole, irradiation, and chemotherapy *without* active immunization in a prior trial [34]. The unusually long survival time observed so far in patients actively immunized with the U-251MG glioma cell line requires confirmation and further investigation in a phase II study to evaluate specifically the response to active immunotherapy and conventional therapy as compared to the response to conventional therapy alone.

Future Directions

The future of immunotherapy in the management of patients with anaplastic gliomas is still undefined. Useful immunological data are now being acquired, and current preliminary studies offer some hope for the development of active immunotherapy as a legitimate therapeutic modality for patients with anaplastic gliomas. If appropriate controls can be applied to obviate allergic encephalomyelitis, it may be possible to evaluate autologous human glioma tissue as an immunogen in the future. The emergence of monoclonal antibody production techniques should afford the opportunity for using these antibodies as reagents in neurodiagnostic studies, and perhaps as therapeutic agents themselves (complement-dependent humoral cytotoxicity) or as carriers of cytotoxic agents in the form of radioactivity or a drug toxic to neoplastic brain cells. The attachment of polyhedral boron to monoclonal antiglioma antibodies may be a means of enhancing radiotherapeutics [3].

Autoadoptive immunotherapy using the patient's own peripheral-blood mononuclear leukocytes awaits the refinement of techniques to specifically activate, clone, and expand antiglioma effector cells that can be safely transferred back to the patient. The systemic and intratumoral means of administering these antitumor cells will have to be compared to establish the risk-benefit ratio. In the meanwhile, it would greatly facilitate the construction of rational therapeutic protocols to have more information regarding the effects of steroids as they may interact with immunotherapeutic treatments. In addition, we must have a current assessment of the relative importance of serum blocking factor(s), helper cells, and suppressor cells as they might influence the treated host's response to an intracranial neoplasm. Much is yet to be learned, and much remains to be gained before it will be possible to control the growth of anaplastic gliomas and their devastating effects on our patients and their families. In the final analysis, even if immunotherapy does not afford a means to resolve the currently dismal prognosis for patients with malignant gliomas, a scientifically sound pursuit of this potential modality for therapy should afford more thorough scrutiny of the biology of CNS neoplasia.

Acknowledgement

This work was supported in part by USPHS research grants CA 29115, CA 29125, CA 30479 and CA 16086 as well as contract CM 07348.

References

1 Apuzzo, M.L.J.; Sheikh, K.M.A.; Weiss, M.H.; Heidne, J.S.; Kurze, T.: The utilization of native glioma antigens in the assessment of specific cellular and humoral immune responses in malignant glioma patients. Acta neurochir. *55:* 180–200 (1981).

2 Arai, T.; Nagai, M.; Kasage, S.; Yasuda, S.: The effect of human fibroblast interferon (Huifn-beta) on malignant tumors; in vitro experiments. Abstract. 22nd Workshop on Selected Topics of the Jap. Soc. Neuropathol., Fukuoka 1981, p. 17.

3 Barth, R.F.; Johnson, C.W.; Wei, W.Z.; Corey, W.E.; Soloway, A.H.; McQuire, J.: Neutron capture using boronated monoclonal antibody directed against tumor associated antigens. Cancer Detect. Prevent. (in press, 1982).

4 Bernstein, M.; Gutin, P.H.: Interstitial irradiation of brain tumors: a review. Neurosurgery *9:* 741–750 (1981).

5 Bigner, D.D.; Bigner, S.H.; Ponten, J.; Westermark, B.; Mahaley, M.S., Jr.; Ruoslahti, E.; Herschman, H.; Eng, L.F.; Wikstrand, C.J.: Heterogeneity of genotypic and phenotypic characteristics of fifteen permanent cell lines derived from human gliomas. J. Neuropath. exp. Neurol. *15:* 201–209 (1981).

6 Bigner, D.D.; Pitts, O.M.; Wikstrand, C.J.: Induction of lethal experimental allergic encephalomyelitis in nonhuman primates and guinea pigs with human glioblastoma multiforme tissue. J. Neurosurg. *55:* 32–42 (1981).

7 Bleehen, N.M.; Wiltshire, C.R.; Plowman, P.N.; Watson, J.V.; Gleave, J.R.; Holmes, A.E.; Lewin, W.S.; Treip, C.S.; Hawkins, T.D.: A randomized study of misonidazole and radiotherapy for grade 3 and 4 cerebral astrocytoma. ICRDB Cancergram, Series CT18, No.81/11, p.2 (US Department of Health and Human Services, 1981).

8 Bloom, W.H.; Carstaris, K.C.; Crompton, M.R.; McKissock, W.: Autologous glioma transplantation. Lancet *ii:* 77–78 (1960).

9 Bloom, H.J.; Peckham, M.J.; Richardson, A.E.; Alexander, P.A.; Payne, P.M.: Glioblastoma multiforme: a controlled trial to assess the value of specific active immunotherapy in patients treated by radical surgery and radiotherapy. Br. J. Cancer *27:* 253–267 (1973).

10 Braun, D.P.; Penn, R.D.; Flannery, A.M.; Harris, J.E.: Immunoregulatory cell function in peripheral blood leukocytes of patients with intracranial gliomas. Neurosurgery *10:* 203–209 (1982).

11 Brooks, W.H.; Latta, R.B.; Mahaley, M.S.; Roszman, T.L.; Dudka, L.; Skaggs, C.: Immunobiology of primary intracranial tumors. Part 5: Correlation of a lymphocyte index and clinical status. J. Neurosurg. *54:* 331–337 (1981).

12 Brooks. W.H.; Netsky, M.G.; Normansell, D.E.; Horowitz, D.A.: Depressed cell-mediated immunity in patients with primary intracranial tumors. J. exp. Med. *136:* 1631–1647 (1972).

13 Cheever, M.A.; Greenberg, P.D.; Fefer, A.; Gillis, S.: Augmentation of the anti-tumor therapeutic efficacy of long-term cultured T lymphocytes by in vivo administration of purified interleukin-2. J. exp. Med. *155:* 968–980 (1982).

14 Chretien, B.P.; Lipson, S.D.; Makuch, R.W.; Kenady, D.E.; Cohen, M.H.: Effects of thymosin in vitro in cancer patients and correlation with clinical course after thymosin immunotherapy. Ann. N. Y. Acad. Sci. *332:* 135–147 (1979).

15 Ciembroniewicz, J.; Kolar, O.: Tissue culture studies of glioblastoma multiforme. Acta cytol. *13:* 42–49 (1969).

16 Coakham, H.B.; Kornblith, P.L.; Quindlen, E.A.; Pollock, L.A.; Wood, W.C.; Hartnett, L.C.: Autologous humoral response to human gliomas and analysis of certain cell surface antigens: in vitro study with the use of microcytotoxicity and immune adherence assays. J. natn. Cancer Inst. *64:* 223–233 (1980).

17 Decarvalho, S.; Kaufman, A.; Pineda, A.: Adjuvant chemo-immunotherapy in central nervous system tumors; in Salmon, Josen, Adjuvant therapy of cancer, pp.495–502 (Elsevier/North Holland, Amsterdam 1977).

18 de Tribolet, N.; Martin-Achard, A.; Louis, J.A.: Circulating immune complexes in patients with gliomas. Acta neurochir. *28:* suppl., pp.437–474 (1979).

19 Egger, A.E.: Autoradiographic and fluorescence antibody studies of the human host immune response to gliomas. Neurology, Minneap. *22:* 246–250 (1972).

20 Febvre, H.; Maunoury, R.; Constans, J.P.; Trouillas, P.: Réactions d'hypersensibilité retardée avec des lignées de cellules tumorales cultivées in vitro chez des malades porteurs de tumeurs cérébrales malignes. Int. J. Cancer *10:* 221–232 (1972).

21 Gillis, S.; Baker, P.E.; Ruscetti, F.W.; Smith, K.A.: Long-term culture of human antigen specific cytotoxic T cell lines. J. exp. Med. *148:* 1093–1098 (1978).

22 Grace, J.T., Jr.; Perese, D.M.; Metzgar, R.S.; Sasabe, T.; Holdridge, B.: Tumor autograft responses in patients with glioblastoma multiforme. J. Neurosurg. *18:* 159–167 (1961).

23 Hashim, G.A.: Myelin basic protein: structure, function and antigenic determination. Immunol. Rev. *39:* 60–107 (1978).

24 Huang, K.Y.; Kind, P.D.; Jagoda, E.M.; Goldstein, A.L.: Thymosin treatment modulates production of interferon. Int. J. Interferon Res. (in press, 1983).

25 Kornblith, P.L.; Dohan, F.C.; Wood, W.; Whitman, B.O.: Human astrocytomas: serum-mediated immunologic response. Cancer *33:* 1512–1519 (1974).

26 Kumar, S.; Taylor, G.; Steward, J.K.; Waghe, M.A.; Morris-Jones, P.: Cell-mediated immunity and blocking factors in patients with tumours of the central nervous system. Int. J. Cancer *12:* 194–205 (1973).

27 Lundblad, D.; Lundgren, E.: Block of a glioma cell line in S by interferon. J. Cancer *27:* 749–754 (1981).

28 Lundblad, D.; Lundgren, E.: Changes in DNA-polymerase and thymidine kinase activity during interferon treatment. Biochem. biophys. Res. Commun. *105:* 1569–1576 (1982).

29 Mahaley, M.S., Jr.; Bigner, D.D.; Dudka, L.F.; Wilds, P.R.; Williams, D.H.; Bouldin, T.W.; Whitaker, J.N.; Bynum, J.M.: Immunobiology of primary intracranial tumors. Part 7: Active immunization of patients with anaplastic human glioma cells. A pilot study. J. Neurosurg. *59:* 201–207 (1983).

30 Mahaley, M.S., Jr.; Gillespie, G.Y.; Gillespie, R.P.; Watkins, P.J.; Bigner, D.D.; Wikstrand, C.J.; Macqueen, J.M.; Sanfilippo, F.: Immunobiology of primary intracranial tumors. Part 8: Serological responses to active immunization of patients with anaplastic gliomas. J. Neurosurg. *59:* 208–216 (1983).

31 Mahaley, M.S., Jr.; Brooks, W.H.; Roszman, T.L.; Bigner, D.D.; Dudka, L.; Richardson, S.: Immunobiology of primary intracranial tumors. I. Studies of the cellular and humoral general immune competence of brain tumor patients. J. Neurosurg. *46:* 467–476 (1977).

32 Mahaley, M.S., Jr.; Gillespie, G.Y.: Immunologic considerations of patients with brain tumors; in Walker, Oncology of the nervous system (Nijhoff, Boston 1982).

33 Mahaley, M.S., Jr.; Mahaley, J.L.; Day, E.D.: The localization of radio-antibodies in human brain tumors. II. Radioautography. Cancer Res. *25:* 779–793 (1965).

34 Mahaley, M.S., Jr.; Steinbok, P.; Aronin, P.; Dudka, L.; Zinn, D.: Immunobiology of primary intracranial tumors. V. Levamisole as an immune stimulant in patients and in the ASV glioma model. J. Neurosurg. *54:* 220–227 (1981).

35 Marshall, G.D., Jr.; Thurman, G.B.; Goldstein, A.L.: Regulation of in vitro generation of cell-mediated cytotoxicity. In vitro induction of suppressor T lymphocytes by thymosin. J. reticuloendoth. Soc. *28:* 141–149 (1980).

36 Martin-Achard, A.; de Tribolet, N.; Louis, J.A.; Zander, E.: Immune complexes associated with brain tumors: correlation with prognosis. Surg. Neurol. *13:* 161–163 (1980).

37 Martin-Achard, A.; Diserens, A.-C.; de Tribolet, N.; Carrel, S.: Evaluation of the humoral response of glioma patients to a possible common tumor-associated antigen(s). Int. J. Cancer *25:* 219–224 (1980).

38 Mazumder, A.; Grimm, E.A.; Eberlein, T.; Rosenberg, S.A.: Phase I clinical trial of infusion into cancer patients of autologous lymphocytes activated in vitro with lectins. Proc. Am. Soc. clin. Oncol. *1:* C158 (1982).

39 McGregor, D.D.; Logie, P.S.: The mediator of cellular immunity. VII. Localization of sensitized lymphocytes in inflammatory exudates. J. exp. Med. *139:* 1415–1430 (1974).

40 Morantz, R.A.; Wood, G.W.; Foster, M.; Clark, M.; Gollahon, K.: Macrophages in experimental and human brain tumors. Part 2: Studies of the macrophage content of human brain tumors. J. Neurosurg. *50:* 305–311 (1979).

41 Nagai, M.; Arai, T.; Ueda, Y.; Kasage, S.; Kono, S.; Kohase, M.: Interferon therapy for malignant brain tumors. Abstract. Proc. 39th Ann. Meet. Japan Neurol. Soc., pp. 269 (Kanazawa, Japan 1980).

42 Neuwelt, E.A.; Clark, K.; Kirkpatrick, J.B.; Toben, H.: Clinical studies of intrathecal autologous lymphocyte infusions in patients with malignant gliomas: a toxicity study. Ann. Neurol. *4:* 307–312 (1978).

43 Ommaya, A.K.; Reed, J.; Walters, C.L.; Meeker, W.R.; Weiss, J.F.: Thymosin for brain tumor therapy. A phase I trial in patients with malignant gliomas. Abstr. AANS Meet. (Poster Section), 1981, p. 90.

44 Quindlen, E.A.; Dohan, F.C.; Kornblith, P.L.: Improved assay for cytotoxic antiglioma antibody. Surg. Forum *25:* 464–466 (1974).

45 Ridley, A.; Cavanagh, J.B.: Lymphocytic infiltration in gliomas: evidence of possible host resistance. Brain *94:* 117–124 (1971).

46 Roszman, T.L.; Brooks, W.H.; Elliott, L.H.: Immunobiology of primary intracranial tumors. VIII. Suppressor cell function and lectin-binding lymphocyte subpopulations in patients with cerebral tumors. Cancer *50:* 1273–1279 (1982).

47 Ruscetti, F.W.; Gallo, R.C.: Human T-lymphocyte growth factor: regulation of growth and function of T lymphocytes. Blood *57:* 379–394 (1981).

48 Salcman, M.; Samaras, G.M.: Hyperthermia for brain tumors: biophysical rationale. Neurosurgery *9:* 327–335 (1981).

49 Salford, L.G.; Stromblad, L.G.; Nordstrom, C.H.; Hornmark-Stenstem, B.; Brandt, L.; Brismar, J.; Brun, A.; Sjogren, H.O.; Flodgran, P.: Intratumoral administration of interferon in malignant gliomas. Acta neurochir. *56:* 130 (1981).

50 Seeger, R.C.; Danon, Y.L.; Rayner, S.A.; Hoover, F.: Definition of a Thy-1 determinant on human neuroblastoma, glioma, sarcoma, and teratoma cells with a monoclonal antibody. J. Immun. *128:* 983–989 (1982).

51 Selker, R.G.; Wolmark, N.; Fisher, B.; Moore, P.: Preliminary observations on the use of *Corynebacterium parvum* in patients with primary intracranial tumors: effect on intracranial pressure. J. surg. Oncol. *10:* 299–303 (1978).

52 Serrou, B.; Cupissol, D.; Caraux, J.; Thierry, C.; Rosenfeld, C.; Goldstein, A.L.: Ability of thymosin to decrease in vivo and in vitro suppressor cell activity in tumor bearing mice and cancer patients. Recent results. Cancer Res. *75:* 110–114 (1980).

53 Strausser, J.L.; Rosenberg, S.A.: In vitro growth of cytotoxic human lymphocytes. I. Growth of cells sensitized in vitro to alloantigens. J. Immun. *121:* 1491–1495 (1978).

54 Takakura, K.; Miki, Y.; Kubo, O.; Ogawa, N.; Matusutani, M.; Sano, K.: Adjuvant immunotherapy for malignant brain tumors. Jap. J. clin. Oncol. *12:* 109–120 (1972).

55 Trouillas, P.; Lapras, C.L.: Immunothérapie active des tumeurs cérébrales. Neuro-Chirugie *16:* 143–170 (1970).
56 Vanky, F.; Gorsky, T.; Gorsky, Y.; Masucci, M.G.; Klein, E.: Lysis of tumor biopsy cells by autologous T lymphocytes activated in mixed culture and propagated with T cell growth factor. J. exp. Med. *155:* 83–95 (1982).
57 Vanky, F.; Klein, E.: Specificity of auto-tumor cytotoxicity exerted by fresh, activated and propagated human T lymphocytes. Int. J. Cancer *29:* 547–553 (1982).
58 Walker, M.D.; Green, S.B.; Byar, D.P.; Alexander, E., Jr.; Batzdorf, U.; Brooks, W.H.; Hunt, W.E.; Maccarty, C.S.; Mahaley, M.S., Jr.; Mealey, J., Jr.; Owens, G.; Ransohoff, J., II; Robertson, J.T.; Shapiro, W.R.; Smith, K.R., Jr.; Wilson, C.B.; Strike, T.A.: Randomized comparisons of radiotherapy and nitrosoureas for the treatment of malignant glioma after surgery. New Engl. J. Med. *303:* 1323–1329 (1980).
59 Wikstrand, C.J.; Bigner, D.D.: Expression of human fetal brain antigens by human tumors of neuroectodermal origin as defined by monoclonal antibodies. Cancer Res. *42:* 267–275 (1982).
60 Witz, I.P.; Hanna, M., Jr.: In situ expression of tumor immunity (Plenum Press, New York 1980).
61 Wood, G.W.; Morantz, R.A.; Tilzer, S.A.; Gollahon, K.A.: Immunoglobulin bound in vivo to Fc receptor-positive cells in human central nervous system tumors. J. natn. Cancer Inst. *64:* 411–418 (1980).
62 Woosley, R.E.; Mahaley, M.S., Jr.; Mahaley, J.L.; Miller, G.M.; Brooks, W.H.: Immunobiology of primary intracranial tumors. 3. Microcytotoxicity assays of specific immune responses of brain tumor patients. J. Neurosurg. *47:* 871–885 (1977).
63 Young, H.F.; Kaplan, A.M.; Regelson, W.: Immunotherapy with autologous white cell infusions ('lymphocytes') in the treatment of recurrent glioblastoma multiforme. Cancer *40:* 1037–1044 (1977).
64 Young, H.F.; Sakalas, R.; Kaplan, A.M.: Inhibition of cell-mediated immunity in patients with brain tumor. Surg. Neurol. *5:* 19–23 (1976).
65 Zarling, J.M.; Bach, F.H.: Continuous culture of T cells cytotoxic for autologous human leukemia cells. Nature, Lond. *280:* 685–688 (1979).

M.S. Mahaley, Jr., MD, PhD, Division of Neurological Surgery, University of North Carolina, Chapel Hill, NC 27514 (USA)

Local Therapy – A. Radiation Therapy

Prog. exp. Tumor Res., vol. 28, pp. 136–151 (Karger, Basel 1984)

Radiation Necrosis

J. R. Fike[a, b, d], *G. E. Sheline*[a], *C. E. Cann*[b], *R. L. Davis*[c, d]

Departments of [a]Radiation Oncology, [b]Radiology, and [c]Pathology (Neuropathology), and [d]The Brain Tumor Research Center of the Department of Neurological Surgery, School of Medicine, University of California, San Francisco, Calif., USA

Introduction

The use of radiation therapy in the management of primary brain tumors is well established [36]. With the exception of postoperative irradiation, no other single or adjuvant therapy has been able to produce significant palliation or cure for patients with these tumors [33]. Tumor doses above 5,000 rad apparently prolong the median life span of patients with malignant gliomas, and *Walker* et al. [39] reported a dose-effect relationship between 5,000, 5,500, and 6,000 rad. However, radiation has the potential to produce severe injury to normal brain tissue in and around the tumor bed [20, 25, 37]. Photon doses below approximately 5,300 rad delivered in a conventional fractionation scheme appear to be tolerated by the normal adult brain, while a dose of 6,000 rad delivered in 30 fractions over 6 weeks approaches the threshold dose for brain necrosis [25, 37]. Because survival estimates for patients with malignant glioma treated with conventional irradiation regimens have been consistently low, more aggressive radiation treatments have been instituted [33]. With the use of higher radiation doses, the likelihood of injury to the normal brain increases.

Risks associated with any given set of radiation factors that may lead to the development of radiation necrosis have not been clearly established [37]. This is the result at least in part of the anecdotal nature of the literature, in which pertinent factors such as total dose at the site of necrosis, size and fraction number, overall treatment time, and irradiated volume are often unclear or omitted [37]. New treatments that include combined modalities [1], interstitial irradiation [16], high linear energy transfer radiation [35] and hyperthermia [34] add other uncertainties to the determination of risk factors. It is clear that adequate, well-designed animal studies are required to elucidate the pathogenic factors, dynamic aspects, and time-dose fractionation relationships involved in the development of late radiation damage. Such studies will play an important role in the establishment of accurate risk estimates.

This chapter will briefly summarize pertinent factors relating to radiation necrosis, including relevant clinical and experimental studies. In addition, the results of recent investigations into the biology of experimental radiation necrosis will be reviewed.

Clinical Experience

A 5% incidence of radiation necrosis has been reported in patients with primary brain or pituitary tumors who received 4,500 rad or more using 180- to 200-rad fractions [25]. Except for that report, it has not been possible to determine the probability that a particular injury will occur for a given time-dose fractionation scheme [37]. This is because selection factors, low autopsy rates, and a lack of knowledge of the population from which patients were accrued have precluded calculation of incidence [37]. However, circumstances in which adverse reactions have occurred are known, and a number of fine reviews and analyses of the literature are available, which address in detail the problem of risk and incidence [14, 20, 25, 37].

The adverse effects of radiation on the human brain generally have been considered as a function of the time of appearance [37]. Acute reactions, presumably the result of edematous changes or transient demyelination, can occur during treatment but usually can be managed with corticosteroids [37]. Early delayed reactions appear a few weeks to months after irradiation and, although they are generally transient, can be severe and in some cases fatal [21]. Early delayed reactions are characterized histopathologically by extensive demyelination, intense glial cell reaction, and minimal vascular changes [21]. Late delayed reactions, particularly radiation necrosis, constitute the major hazard to brain tissues.

Occurring months to years after exposure, late delayed reactions are generally irreversible, progressive, and dose-related [37]. Rarely observed below 5,200 rad given in 200-rad fractions in 5 weeks, radiation necrosis has been observed with conventionally fractioned total doses of 5,000–6,000 rad [25, 37]. Local tumor doses or boosts as high as

7,500–8,000 rad have been delivered although it was considered probable that these doses would lead to radiation necrosis in some instances [33]. Cyclotron-produced neutrons, although tumoricidal in some patients, have produced extensive and sometimes fatal radiation necrosis [35], possibly because the relative biological effectiveness for fractionated neutron irradiation has been underestimated [18].

Radiation necrosis can occur in most regions of the brain and tends to involve mainly the white matter [1, 15, 24, 26, 37]. Microscopically there are several striking alterations, the interrelationships and pathogenesis of which are unknown or speculative [15, 26]. Vascular change is the most characteristic feature, and thickened vessel walls, fibrinoid necrosis, amyloid and fatty degeneration, endothelial proliferation, perivascular fibrosis, and thrombosis have been observed [1, 15, 23, 24, 26]. Vascular features are not necessarily the primary pathogenic factors. Demyelination, either focal or diffuse, is common and it has been proposed that myelin-producing oligodendrocytes are constitutionally more radiosensitive than other glial cells [24] or are subject to an autoimmune demyelinating process [22]. Another important factor that may potentiate the effects of radiation is cerebral edema [1].

Radiation necrosis presents difficult diagnostic problems because the signs and symptoms often mimic recurrent tumor. Diagnosis by radiologic techniques depends largely upon computed tomography (CT), which can effectively demonstrate the presence of radiation-induced changes such as low density foci, increased mass effects, increased contrast enhancement, and ventricular enlargement [5, 28, 29, 31]. Clinical studies have shown that images of radiation lesions on CT scans correlate well with volumes of brain receiving a threshold dose, determined from isodose reconstruction maps [29]. It has been reported that with increasing radiation dose, contrast enhancement of lesions is progressively more intense [5]. Although CT scans may suggest radiation-induced changes, in current practice the diagnosis should be confirmed by biopsy if the affected volume is near a tumor site.

Under certain conditions, corticosteroids and/or surgical resection may ameliorate or relieve the symptoms of radiation-induced damage to normal tissue [6]. Based on a review of the limited literature, it has been concluded that surgical resection of areas of focal necrosis was often lifesaving and that corticosteroids were effective adjuvants to surgical therapy [6]. In addition, under certain circumstances steroids were effective as a primary therapy. In many patients, however, the necrosis was too extensive or morphologically inaccessible for surgical resection to be of value [6]. Early diagnosis and the ability to differentiate radiation damage from recurrent tumor should facilitate patient management.

Experimental Studies

Nonhuman primate models have been used to investigate the pathologic basis of late radiation damage, qualitatively characterizing the morphologic and physiologic factors involved [2, 3, 19, 38]. Regional and whole brain irradiation using either single doses or fractionated treatments to simulate the clinical situation have been carried out to compare late effects developing under a variety of irradiation conditions. These studies have indicated that the extent of late radiation damage varied with the age of the host, dose and mode of delivery, and the time interval following exposure. Fractionated treatments (200 rad/day, 5 days/week) of 6,000 and 8,000 rad produced lesions very much like those observed 26 weeks after a single 1,500-rad dose, but at 1 year lesions produced by 6,000 rad were quiescent, while the other two doses had produced extensive necrosis [3]. The morphologic appearance of radiation-induced lesions was not unlike that observed in humans, with the hallmark lesion being necrotic foci scattered within the white matter of the brain. Edema, gliosis, impaired blood flow, vascular hyperplasia, and breaks in the blood-brain barrier were observed.

Studies of the effects of radiation on the normal dog brain have indicated that the development, pathologic characterization, and dose requirements for radiation brain necrosis are similar to those observed in nonhuman primates and humans [4, 10, 41]. Evidence of vascular injury as well as possible atrophic changes and ventricular enlargement have been observed in dogs after single or fractionated treatments [10, 41]. Noninvasive CT methods to evaluate radiation damage in the dog brain are now available and sequential appraisal of the development of radiation lesions can be obtained [10].

Brain injury caused by radiation has also been studied in rodents and other animals [17, 18, 27, 30, 32, 40]; the available information suggests that rodent models may be qualitatively and quantitatively different from dogs, nonhuman primates, and humans. Functional impairment of the vasculature in the rat brain was observed only after single doses of 2,000–3,000 rad while at higher doses vascular lesions were reduced and did not contribute to death [30]. After a single 2,000-rad dose to rat brain, a latency period of 1 year or more was required before vascular effects were apparent histologically [32]. Morphologic alterations in the mouse brain were shown to be qualitatively quite different from those observed in nonhuman primates and humans [40].

Recent Advances in the Investigation of Radiation-Induced Brain Damage

CT has been used to follow the response of tumor and normal tissues after therapeutic irradiation [5, 28, 29, 31], but has not been used to characterize the dynamics of the response. The advantage of serial CT studies lies primarily in the capability to monitor the development of changes both qualitatively and quantitatively. The information obtained can then be correlated with physical, neurologic, and pathologic information. A CT image is merely a two-dimensional representation of a three-dimensional structure in a particular region of the body. This information is stored in a digital form and can be accessed to provide quantitative data about the tissue characteristics; this process is called quantitative CT (QCT). The number of discrete changes measurable by QCT provides a large and sensitive data base concerning normal and pathologic morphology and physiology; QCT can be performed repeatedly with minimal risk to the subject. The precision and reproducibility of QCT measurements and the capability to obtain serial measurements require the use of fewer studies to demonstrate quantitatively a change from baseline studies [8].

The canine brain is a good model for radiation-induced brain injury [10, 41], and with QCT we have been studying the development of radiation damage using a canine model. The results we have obtained, reviewed here, illustrate the usefulness of these methods for the study of basic mechanisms and for application to the problem of radiation-induced damage.

Hemibrain X-Irradiation

Single doses of 1,500 or 3,000 rad of megavoltage X-rays were delivered to the right hemisphere of the brain in a group of adult beagle dogs [11]. Doses were chosen to insure the development of radiation-induced changes, and all dogs that received 3,000 rad died within the first 6 months after irradiation, but only 2 of 8 dogs that received 1,500 rad died within

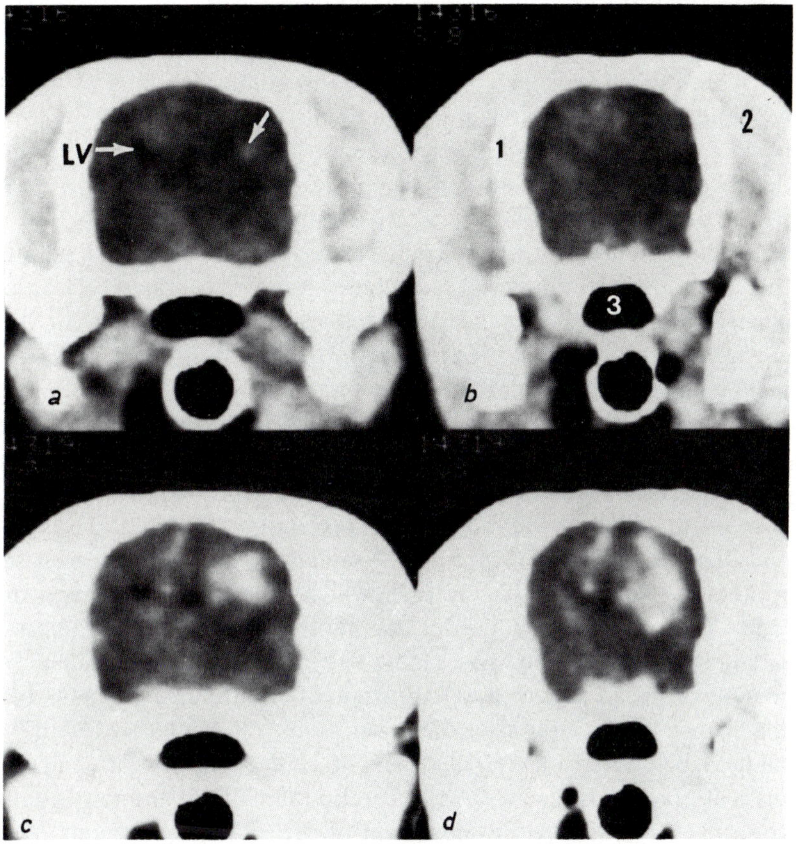

Fig. 1. Representative coronal CT scans of the canine brain 5.5 months after a single dose of 1,500 rad to the right hemisphere, at the level of the thalamus (*a, c*) and the optic chiasm (*b, d*). Scans were obtained before (top) and during (bottom) an infusion of iodinated contrast material. Scans are viewed looking rostrally with the left side of the dog corresponding to the viewer's left. Precontrast scans show displacement of the lateral ventricles (LV), a small focus of increased density (arrow) surrounded by radiolucent material consistent with edema and/or necrosis. Postcontrast scans show a sharply delineated mass characterized by a rather uniform enhancement. Displacement of the falx cerebri and ventricles are clearly shown. The mass is primarily within the white matter, although some extension into cortical gray matter can be seen. Other structures include: 1 = calvaria; 2 = temporalis muscle; 3 = nasopharyngeal canal.

that period. Both qualitative and quantitative differences were observed between the two dose groups.

Densitometric values of various morphologic regions of the brain were measured before and after the administration of contrast material. In normal control dogs, high precision was obtained for precontrast densitometry, both within and among dogs, evaluated over a 9-month period; regional differences of 2–3 Hounsfield units (0.2–0.3%) were statistically significant [8]. In irradiated dogs, decreases in CT number of the cerebrum were observed 3–4 months after irradiation [11]. Although changes were observed in the caudal fossa and midbrain, the most obvious changes were primarily associated with the cerebral white matter, were independent of ventricular enlargement and were, for the most part, progressive with time. The lucent changes primarily reflected edematous and/or necrotic changes and were limited to the irradiated hemisphere; the control or nonirradiated side remained normal.

2–4 weeks after the reduction in precontrast CT number, an abrupt and significant increase in contrast enhancement was observed in the irradiated hemisphere (fig. 1) [11], which presumably reflects a disturbance in the blood-brain barrier that allows contrast medium to accumulate in the extravascular spaces [13]. The increased contrast enhancement was observed after both doses of irradiation, although the magnitude of the change was greater after the lower dose. Initially located in the white matter, regions of contrast enhancement were associated with a significant mass effect that displaced the falx cerebri, and lateral and third ventricles (fig. 1). Histopathologic examination of the enhancing regions revealed pronounced vascular changes in association with areas of coagulation necrosis. Fibrinoid necrosis of vessel walls, endothelial proliferation, and petechial hemorrhages were observed. Noninflammatory in nature, the lesions showed marked gliosis, swelling of axis cylinders and rarefaction of myelin. The changes observed in the dog brain 5–8 months after irradiation are identical to changes found in those humans in whom radiation damage occurs after radiation therapy.

The sequence of events that led to the development of a radiation lesion after 1,500 rad to the dog brain is shown qualitatively and quantitatively in figures 2 and 3. In this dog, CT images ranged in appearance from essentially normal 3.5 months after irradiation, through 2–4 months of marked contrast enhancement and mass effect to a reduction in enhancement at 8 months, and finally to virtually no enhancement 12 months after irradiation. It is clear that interpretation of the appearance of radiation-

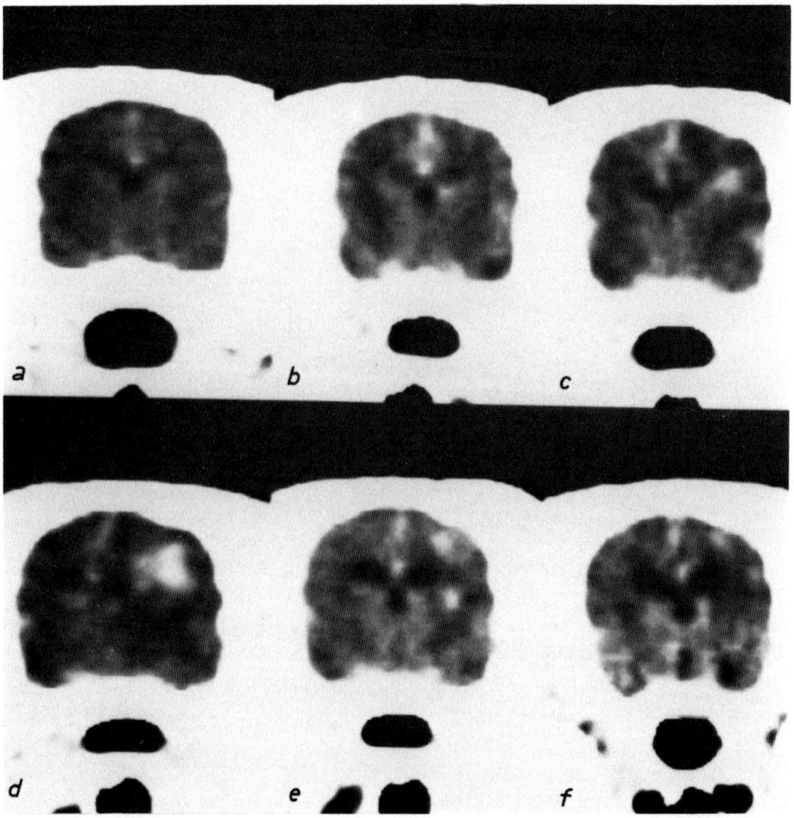

Fig. 2. Coronal CT scans taken at the level of the thalamus during an infusion of contrast material. Scans were taken from a single dog before *(a)*, 3.0 *(b)*, 4.5 *(c)*, 5.5 *(d)*, 8.0 *(e)*, and 12 *(f)* months after a single dose of 1,500 rad to the right hemisphere. Focal enhancement is apparent 4.5 months after irradiation, is maximum at 5.5 months and is markedly reduced by 8 months. At 12 months there is no apparent enhancement, the lateral ventricles are enlarged relative to baseline, and there are some radiolucent changes laterally on the right side.

induced brain damage on a postirradiation CT scan will depend on the time after irradiation. Another important point is that severe vascular-related changes observed in these studies are not necessarily lethal types of changes. The dog survived and at 24 months after irradiation has very minimal neurologic deficits. The fact that dogs that received 3,000 rad showed pronounced ventricular enlargement but minimal contrast

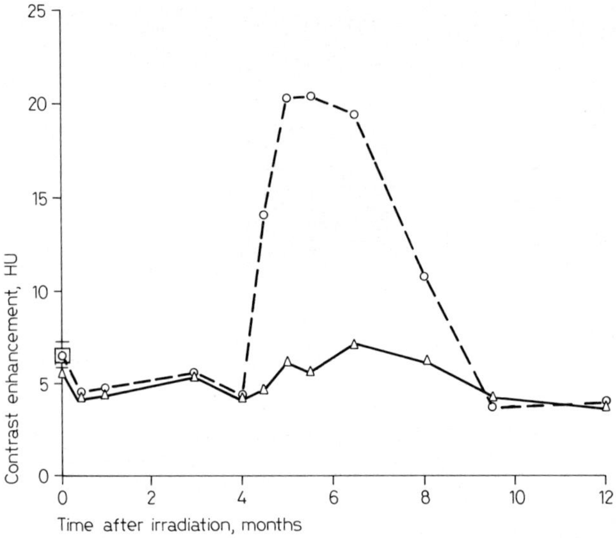

Fig. 3. Contrast enhancement as a function of time after irradiation for the dog in figure 2. Large regions of interest were used to determine CT values, and both irradiated (○) and nonirradiated (△) hemispheres are shown. The square symbol indicates the mean and standard error of the mean for contrast enhancement of the cerebrum of 10 normal adult beagle dogs [8].

enhancement suggests that gross vascular changes may not be the sole factor in radiation-induced lethality at the doses used [11].

Interstitial Irradiation

In conjunction with clinical trials using interstitial ^{125}I for the treatment of malignant glioma [16], we are studying the effects of ^{125}I on the normal brain [9]. Results of preliminary QCT and histopathologic studies suggest that the canine model can be used to measure the effects of interstitial irradiation on normal brain. Removable sources are implanted into the frontal/parietal region of the brain for 1–10 days. Doses 1 cm from the source range from 1,000 to 10,000 rad, with much higher doses nearer the source.

In all dogs studied to date, CT values determined near the source (i.e., high dose) have been significantly reduced relative to normal baseline values. This low density, which develops within days of implantation, has been described histologically as a region of noninflammatory

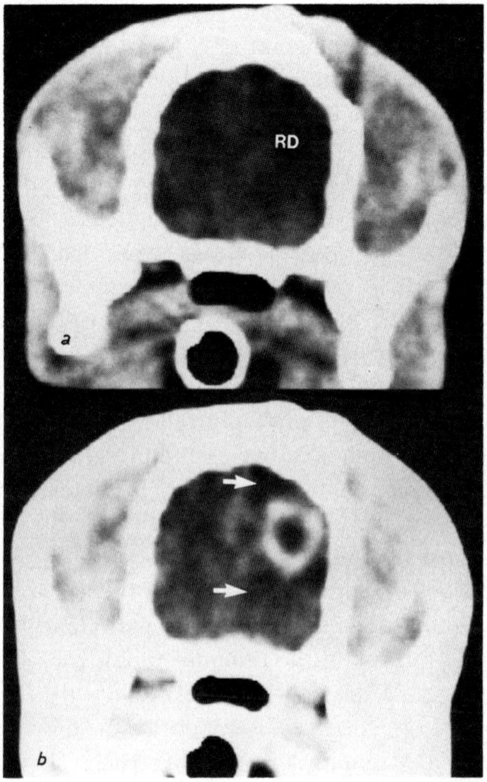

Fig. 4. Coronal CT scans at the level of the thalamus taken 1 month after implantation of an ^{125}I source. Calculated dose at 1 cm was 3,000 rad. Precontrast scan (*a*) shows a large region of reduced density (RD) in the right hemisphere. During an infusion of contrast material (*b*) a very distinct ring of enhancement can be seen. The radiolucent area within the ring is the site of implantation, and corresponds to a region of very high dose. Edema can be observed outside of the ring as a diffuse area of low density (arrows).

coagulation necrosis. At progressively increasing distance from the source, the CT values appear to be inversely related to dose. Contrast enhancement shows a characteristic pattern in all dogs, which consists of a ring of enhancement surrounding the necrotic central region (fig. 4). Distal to the ring of enhancement there is apparent interstitial edema that is transient and resolves 25–50 days after implantation. Quantitatively, the contrast enhancement value for each ring of enhancement appears to increase with time, although the significance of this finding is unclear. In dogs that have survived for more than 2–3 months, the ring of enhancement appears to be reduced in size and amount of contrast enhancement, presumably because of a resolution of the gross vascular manifestation of radiation damage, and is similar to the type of response seen in figure 3.

Pathologically, the changes observed after ^{125}I irradiation are very similar to those observed after external beam X-irradiation, except that in the former the damage is much more localized; there is a focus of coagulation necrosis that is surrounded by presumably viable tissue, which shows the vascular alterations described above. Histologic changes induced by megavoltage X-irradiation become apparent approximately 5 months after irradiation, but similar types of change caused by ^{125}I X-rays appear as early as 1–2 months after implantation. The apparent acceleration in the development of radiation damage is probably related to total dose, dose rate, total time of irradiation, and irradiated volume. At present the importance and interrelationship of the various factors is unclear.

Contrast Accumulation and Washout

The differentiation between radiation necrosis and recurrent tumor cannot be made on CT with any degree of certainty, particularly if the radiation damage is located at the site of the previous tumor [5]. The CT appearance of radiation damage in the dog brain, especially the intense contrast enhancement and mass effect (fig. 1), is not unlike the CT appearance of spontaneously developed gliomas in the dog [12]. Therefore, ongoing studies of the kinetics of contrast uptake and washout are being conducted [7] in an effort to differentiate tumor from radiation damage.

In the evaluation of the irradiated canine brain, contrast enhancement, as determined by differences in CT numbers, is determined from scans obtained before and during an infusion of contrast material [8]. After completion of the latter scans, the infusion is terminated and scans identical to the infusion series are obtained at times from 5 min to 1.5 h after infusion. There is a marked difference in terms of contrast uptake between the irradiated region, where peak enhancement occurs after the infusion, and the nonirradiated side, where peak contrast enhancement occurs during the infusion (fig. 5). To date, all dogs, whether they received X-irradiation or ^{125}I irradiation, have shown the same basic pattern. In contrast, a limited number of spontaneous canine gliomas have shown a very different response, in which maximum contrast enhancement is reached during the infusion [7] (fig. 5). These studies must be repeated in many more irradiated and tumor-bearing dogs before absolute conclu-

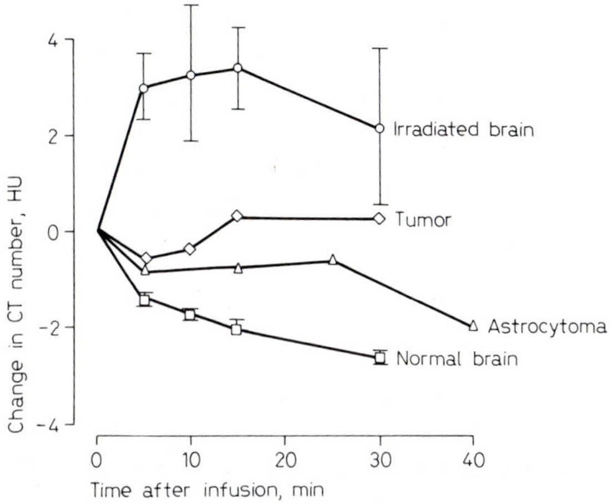

Fig. 5. Accumulation and washout of contrast medium from irradiated normal brain and from spontaneous brain tumors in dogs. Symbols for irradiated (○) and normal, nonirradiated (□) brain represent the means and standard errors for 5 dogs that received a single 1,500-rad dose to one hemisphere of the brain. The two tumors shown, one cerebellar ('tumor') and the other a midbrain astrocytoma, are representative of the type of response observed in 5 tumor-bearing dogs, and are clearly different from that in irradiated normal brain.

sions can be made; however, the preliminary results indicate that contrast uptake may be inherently different between tumor and regions of radiation damage. If so, this type of study could have a significant impact in the diagnosis of radiation-induced brain damage.

Advanced Diagnostic Techniques

While QCT methods for analysis of serial changes in the brain are just becoming more widely used, even newer techniques for noninvasive monitoring of radiation-induced changes are becoming available. Rapid sequential CT scanning of a region of the brain during and immediately after the bolus injection of contrast material can be used to study the short-term kinetics of contrast passage; these results can be related to the transit time and volume of blood in the tissue. The current generation of scanners can provide this dynamic scanning with about 1 image every 2–3 s; newer scanners being developed will be able to scan the whole brain in 200 ms with a repetition rate of once per second, or to scan a smaller region every 50 ms, which will provide true blood flow data. QCT with dual-energy scanning can give information about tissue chemistry characteristics. Nuclear magnetic resonance also has the potential for true chemical characterization of tissues and may provide the best method for the study of such processes as demyelination and transient edema. All of these newer techniques will be useful in the serial evaluation of brain response to radiation; it only remains for the proper measurements and methods to be identified and developed.

Conclusions

The use of more aggressive therapeutic approaches in the management of malignant gliomas may increase the likelihood of radiation-induced damage to normal tissue. The clinical management of radiation necrosis may depend to some extent on the knowledge of its morphologic and physiologic development. The studies discussed here use the unique capabilities of QCT for the noninvasive evaluation of radiation effects and may ultimately define some of the factors involved in radiation lesions, including time-dose relationships, effect of dose rate and irradiated volume. In addition, new diagnostic techniques that may aid in the identification of radiation changes and that may be instrumental in their treatment are being developed.

Acknowledgments

The authors thank Drs. *Theodore L. Phillips* and *Steven A. Leibel* for their comments and criticisms. Supported in part by NIH Program Project Grant CA-13525, NIH Grant CA-30445, and funds from the University of California Cancer Research Coordinating Committee. We thank *Neil Buckley* for editorial assistance.

References

1 Burger, P.C.; Mahaley, M.S., Jr.; Dudka, L.; Vogel, F.S.: The morphologic effects of radiation administered therapeutically for intracranial gliomas. A postmortem study of 25 cases. Cancer *44:* 1256–1272 (1979).

2 Caveness, W.F.: Pathology of radiation damage to the normal brain of the monkey. Natn. Cancer Inst. Monogr. *46:* 57–76 (1977).

3 Caveness, W.F.: Experimental observations; delayed necrosis in normal monkey

brain; in Gilbert, Kagan, Radiation damage to the nervous system. A delayed therapeutic hazard; 1st ed., pp. 1–38 (Raven Press, New York 1980).

4 Csanda, E.: Early and late brain edema following focal high dose irradiation. Stereotactic cerebral irradiation; in Szikla, Inserm Symp. No. 12, pp. 137–147 (Elsevier/North Holland, Amsterdam 1979).

5 Deck, M. D. F.: Imaging techniques in the diagnosis of radiation damage to the nervous system; in Gilbert, Kagan, Radiation damage to the nervous system. A delayed therapeutic hazard; 1st ed., pp. 107–127 (Raven Press, New York 1980).

6 Edwards, M. S.; Wilson, C. B.: Treatment of radiation necrosis; in Gilbert, Kagan, Radiation damage to the nervous system. A delayed therapeutic hazard; 1st ed., pp. 129–144 (Raven Press, New York 1980).

7 Fike, J. R.; Cann, C. E.: Kinetics of contrast accumulation and washout in canine brain tumors and irradiated normal brain using CT. Radiology (in press, 1984).

8 Fike, J. R.; Cann, C. E.; Berninger, W. H.: Quantitative evaluation of the canine brain using computed tomography. J. Comput. Assist. Tomogr. *6:* 325–333 (1982).

9 Fike, J. R.; Cann, C. E.; Bernstein, M.; Davis, R. L.; Gutin, P. H.; Phillips, T. L.; Weaver, K.: Radiation damage to the canine brain after ^{125}I implantation (Abstract). Radiat. Res. *91:* 370 (1982).

10 Fike, J. R.; Cann, C. E.; Davis, R. L.; Phillips, T. L.: Radiation effects in the canine brain evaluated by quantitative computed tomography. Radiology *144:* 603–608 (1982).

11 Fike, J. R.; Cann, C. E.: Radiation effects in the canine brain quantified by CT. J. Comput. Assist. Tomogr. (Abstract). *7:* 565–566 (1983).

12 Fike, J. R.; Le Couteur, R. A.; Cann, C. E.; Pflugfelder, C. M.: Computerized tomography of brain tumors of the rostral and middle fossas in the dog. Am. J. vet. Res. *42:* 275–281 (1981).

13 Gado, H. H.; Phelps, M. E.; Coleman, R. E.: An extravascular component of contrast enhancement in cranial computed tomography. Part II. Contrast enhancement and the blood-brain barrier. Radiology *117:* 595–597 (1975).

14 Gilbert, H. A.; Kagan, A. R.: Radiation damage to the nervous system. A delayed therapeutic hazard; 1st ed., pp. 1–209 (Raven Press, New York 1980).

15 Groothuis, D. R.; Vick, N. A.: Radionecrosis of the central nervous system: the perspective of the clinical neurologist and neuropathologist; in Gilbert, Kagan, Radiation damage to the central nervous system. A delayed therapeutic hazard; 1st ed., pp. 93–106 (Raven Press, New York 1980).

16 Gutin, P. H.; Phillips, T. L.; Hosobuchi, Y.; Wara, W. M.; Mackay, A. R.; Weaver, K.; Lamb, S.; Hurst, S.: Permanent and removable implants for the brachytherapy of brain tumors. Int. J. Radiat. Oncol. Biol. Phys. *7:* 1371–1381 (1981).

17 Hornsey, S.; Morris, C. C.; Myers, R.: The relationship between fractionation and total dose for x-ray induced brain damage. Int. J. Radiat. Oncol. Biol. Phys. *7:* 393–396 (1981).

18 Hornsey, S.; Morris, C. C.; Myers, R.; White, A.: Relative biological effectiveness for damage to the central nervous system by neutrons. Int. J. Radiat. Oncol. Biol. Phys. *7:* 185–189 (1981).

19 Kemper, T. L.; O'Neill, R.; Caveness, W. F.: Effects of single dose supervoltage whole brain radiation in *Macaca mulatta.* J. Neuropath. exp. Neurol. *37:* 916–940 (1977).

20 Kramer, S.: The hazards of therapeutic irradiation of the central nervous system. Clin. Neurosurg. *15:* 301–318 (1968).

21 Lampert, P.W.; Davis, R.L.: Delayed effects of radiation on the human central nervous system. 'Early' and 'late' delayed reactions. Neurology *14:* 912–917 (1964).

22 Lampert, P.; Tom, M.I.; Rider, W.D.: Disseminated demyelination of the brain following CO^{60} (gamma) radiation. Archs Path. *68:* 322–330 (1959).

23 Mandybur, T.I.; Gore, I.: Amyloid in late postirradiation necrosis of brain. Neurology *19:* 983–992 (1969).

24 Manz, H.J.; Woolley, P.V., III; Ornitz, R.D.: Delayed radiation necrosis of brainstem related to fast neutron irradiation. Case report and literature review. Cancer *44:* 473–479 (1979).

25 Marks, J.E.; Baglan, R.J.; Prassad, S.C.; Blank, W.F.: Cerebral radionecrosis: incidence and risk in relation to dose, time, fractionation and volume. Int. J. Radiat. Oncol. Biol. Phys. *7:* 243–252 (1982).

26 Martins, A.N.; Johnston, J.S.; Henry, J.M.; Stoffel, T.J.; Dichiro, G.: Delayed radiation necrosis of the brain. J. Neurosurg. *47:* 336–345 (1977).

27 McDonald, L.W.; Hayes, T.L.: The role of capillaries in the pathogenesis of delayed radionecrosis of brain. Am. J. Path. *50:* 745–764 (1967).

28 Mikhael, M.A.: Radiation necrosis of the brain: correlation between computed tomography, pathology and dose distribution. J. Comput. Assist. Tomogr. *2:* 71–80 (1978).

29 Mikhael, M.A.: Radiation necrosis of the brain: correlation between patterns on computed tomography and dose of radiation. J. Comput. Assist. Tomogr. *3:* 241–249 (1979).

30 Moustafa, H.F.; Hopewell, J.W.: Late functional changes in the vasculature of the rat brain after local irradition. Br.J. Radiol. *53:* 21–25 (1980).

31 Pay, N.T.: Carella, R.J.; Lin, J.P.: Kricheff, I.I.: The usefulness of computed tomography during and after radiation therapy in patients with brain tumors. Radiology *121:* 79–83 (1976).

32 Reinhold, H.S.; Hopewell, J.W.: Late changes in the architecture of blood vessels of the rat brain after irradiation. Br.J. Radiol. *53:* 693–696 (1980).

33 Salazar, O.M.; Rubin, P.; Feldstein, M.L.; Pizzutiello, R.: High dose radiation therapy in the treatment of malignant gliomas: final report. Int. J. Radiat. Oncol. Biol. Phys. *5:* 1733–1740 (1979).

34 Salcman, M.; Samaras, G.M.: Hyperthermia for brain tumors; biophysical rationale. Neurosurgery *9:* 327–335 (1981).

35 Shaw, C.M.; Sumi, S.M.; Aovord, E.C.; Gerdes, A.J.; Spence, A.; Parker, R.G.: Fast-neutron irradiation of glioblastoma multiforme. Neuropathological analysis. J. Neurosurg. *49:* 1–12 (1978).

36 Sheline, G.E.: Radiation therapy of tumors. Cancer *39:* 873–881 (1977).

37 Sheline, G.E.; Wara, W.M.; Smith, V.: Therapeutic irradiation and brain injury. Int. J. Radiat. Oncol. Biol. Phys. *6:* 1215–1228 (1980).

38 Wakisaka, S.; O'Neill, R.R.; Kemper, T.L.; Verrelli, D.M.; Caveness, W.F.: Delayed brain damage in adult monkeys from radiation in the therapeutic range. Radiat. Res. *80:* 277–292 (1979).

39 Walker, M.D.; Strike, T.A.; Sheline, G.E.: An analysis of dose-effect relationship in the radiotherapy of malignant gliomas. Int. J. Radiat. Oncol. Biol. Phys. *5:* 1725–1731 (1979).

40 Yoshii, Y.; Phillips, T.L.: Late effect of whole brain x-irradiation in LAF_1 male mice. (Abstract) Radiat. Res. *83:* 378–379 (1980).
41 Zook, B.C.; Bradley, E.W.; Casarett, G.W.; Rogers, C.C.: Pathologic findings in canine brain irradiated with fractionated fast neutrons or photons. Radiat. Res. *89:* 562–578 (1980).

J.R. Fike, PhD, Departments of Radiation Oncology and Radiology,
and The Brain Tumor Research Center of the Department of Neurological Surgery,
University of California, San Francisco, CA 94143 (USA)

Prog. exp. Tumor Res., vol. 28, pp. 152–165 (Karger, Basel 1984)

Novel Fractionation Schemes and High Linear Energy Transfer[1]

B. G. Douglas[a], *J. R. Castro*[b]

[a] British Columbia Cancer Research Center and Cancer Control Agency of British Columbia, Vancouver, B.C., Canada; [b] Division of Biology and Medicine, University of California Lawrence Berkeley Laboratory, Calif.; [b] Department of Radiation Oncology, School of Medicine, University of California, Berkeley, Calif., USA

Superfractionation

In the past, improvements in radiation therapy have evolved slowly from empirically based alterations in either the overall time of irradiation or the fractionation of the dosage. Superfractionation is the term used for a recently developed form of radiation therapy in which more than 1 treatment – each treatment being smaller than a conventional dose fraction – is given per day to the patient. In contrast to previous changes in radiation therapy, superfractionation has been developed because there were indications from experimental radiobiologic studies that the use of multiple small doses of radiation might result in significantly improved therapeutic results. The drive to investigate superfractionation resulted directly from a better understanding of how irradiation works, and attempts to exploit the differences between tumors and normal tissues to the patient's advantage.

In therapy with any form of radiation, the total dosage that can safely be given to the patient varies depending on both the number of dose fractions and the time intervals between them. The available clinical experience suggests that a formula expressing mathematically the maximum total dosage, the TD_{max} formula [16], will provide a guideline for equating the injury to normal tissue between daily and superfractionation treatment schedules. Future clinical experience should provide evidence to validate or suggest modifications of the TD_{max} formula.

[1] Supported in part by NCI CA 19138, 15184, 21744 and DOE DE-AC03-76SF00098.

Theory of Superfractionation

Superfractionation can be defined as the use of 2 or more radiation treatments, each with less than 1.02 Gy/day. The use of superfractionation and the rationale behind it are a recent outgrowth of radiobiologic research [6, 17, 38]. In clinical practice, the dose of radiation that can be administered is limited by the damage that can be tolerated by the normal tissues that must be included in the treatment volume. These are generally slowly proliferating tissues, such as fibrovascular connective tissues, or in the case of the brain, the glial tissues. Compared to these tissues, tumors proliferate rapidly. Although this means that, in general, tumors have a proliferative advantage, there are ways in which their associated progression through the cell cycle can be turned against them. The sensitivity of cells to the killing effects of radiation is not uniform throughout their cell cycle; therefore, the effect of irradiation is not uniform throughout the cell cycle. The effect of a dosage of radiation is to kill those cells that are in the more sensitive phases of the cell cycle. Thus, the cells that remain after receiving a radiation dosage are mainly those that were in the most radiation-resistant phases of the cell cycle. Any progression through the cell cycle leads them to phases of the cycle during which they are more sensitive. *Withers* [38] suggested that the use of superfractionation treatment would increase the likelihood that each tumor cell would be in a sensitive phase of the cell cycle during the administration of 1 of the dose fractions, thus increasing the likelihood that the cell would not survive treatment. As the dose-limiting cells of normal tissues are essentially noncycling, and thus are not susceptible to this effect, superfractionation has the advantage over conventional radiation treatment.

There is also evidence that superfractionation will be advantageous because of factors inherent in the shape of the cell-survival curve. Recent radiobiologic studies have indicated that, in situations in which the response of the tissue is dominated by well-oxygenated cells, the curves for cell survival in vivo after irradiation have a shape that can be described by a quadratic cell survival equation [16, 19, 25] of the form:

$$\log S = \alpha D + \beta D^2, \qquad (1)$$

in which S is the surviving fraction of cells, and α and β are cell-survival parameters that are peculiar to the cell line in question. Equation 1 can also be written:

$$(\log S)/\alpha = D + \beta/\alpha\ D^2. \qquad (2)$$

In general, the exact values for α and β are not available from in vivo studies, but in vivo isoeffect studies, in which (log S/α) remains constant, can be used to provide a value for the ratio β/α. The relative values for this ratio for the response-limiting tumor and normal tissues determine what dosage per fraction would be optimal in any individual treatment situation. In any situation in which the tumor is dominated by well-oxygenated cells and in which the β/α ratio is less for the tumor tissue than for the dosage-limiting normal tissue, it is anticipated that superfractionation will provide a therapeutic advantage over conventional dose fractionation. This appears to be the case for most tumors, including malignant brain tumors [7, 17, 18]. When the reverse is true – when the tumor β/α value is greater than that of the dosage-limiting normal tissue – a small number of large dose fractions would produce the best results, provided that hypoxic cells do not inhibit the response of the tumor to the treatment.

There are commonly present in tumor masses areas where the cells are viable but hypoxic. Hypoxic cells are more resistant to the effects of irradiation than are well-oxygenated cells, and although reoxygenation does appear to take place during treatment, the hypoxic cells may determine the response of many tumors to treatment. When the tumor is dominated by hypoxic cells, it is anticipated that superfractionation will produce an even greater improvement in results than in the cases in which reoxygenation is sufficient to the extent that the response is determined by well-oxygenated cells.

The factor by which a dose of radiation must be increased to kill cells that have been made hypoxic is known as the oxygen enhancement ratio (OER). If the OER is constant over the clinically relevant range, superfractionation would be anticipated to provide improved results against any tumor with a β/α value of less than the OER times the β/α value for the dose-limiting normal tissue [17]. From the information that is currently available, it seems likely that any tumor with a response dominated by hypoxic cells would respond better to superfractionation than to conventional radiation treatment. In addition, experiments designed to determine the value of the OER for the radiation doses used in clinical practice are indicating that the OER is less for small radiation doses than for large doses [29]. If this is true in the clinical setting, then the smaller superfractionation doses will provide an even greater improvement in results [17]. Thus, for tumors in which hypoxia is a problem, theory suggests that superfractionation will provide a major improvement in results.

One final factor will be of importance in determining the results of treatment with superfractionation. When tissues are irradiated, 3 different things may happen to their cells. First, some of the cells may totally escape damage, in which case they survive until the next irradiation. Second, some cells receive so much damage that they die either at the first attempt at cell division or after a few cell divisions. The third possibility is that some of the cells are damaged by irradiation to a level insufficient to kill them and, over a period of time, repair the damage; given sufficient time, these cells return to a normal, nondamaged state. This repair is referred to as the repair of sublethal damage.

In vivo, the rates of repair of sublethal injury are exponential in nature, the half-time for repair for normal skin in vivo being about 2.5 h [16]. With conventional radiation therapy, the time interval between treatments is long compared to the time needed for repair of sublethal radiation damage for well-oxygenated tissues, and probably for most tumor tissues as well. When superfractionation is used, however, the time between treatment is such that the repair of sublethal damage from the previous dose of radiation is incomplete when a new dose is given. Therefore, cells that have not fully repaired the damage from irradiation received previously are more susceptible to the next dosage. The new damage, which may not itself be sufficient to kill the cell, adds to the currently existing damage that would have been repaired had the cell had enough time. The additive effect of the 2 dosages together may result in cell death. It can be seen that if there is a big difference in the rates of repair of sublethal injury, then the use of superfractionation will result in a relative decrease in the number of cells that survive among those cells that are slowest in the repair process. At present, there is very little information available on the in vivo rates of repair of sublethal injury in either normal or tumor tissues. There is evidence, however, that – in contrast to more oxygenated cells – hypoxic cells are slow to repair sublethal damage [24]. Superfractionation schedules might provide a means for exploiting this differential rate of repair of sublethal damage between the putatively response-determining hypoxic tumor cells and the more well-oxygenated normal tissues [17].

Clinical Results with Superfractionation

As mentioned earlier, when a new radiation fractionation scheme is to be investigated in the clinic, care must be exercised to ensure, as far as possible, that the doses used in the arms of the study produce equivalent effects in normal tissue. The results of a phase I–phase II study, a sequential nonrandomized study with a sizable historical control group, have been reported [18]. This study was done for 2 reasons: (1) to define the total superfractionated dose equivalent to 40 Gy in 4 weeks of daily treatment, and (2) to see if superfractionation held promise for the treatment of patients with malignant brain tumors. The results obtained with superfractionation were superior to the result in the historical control group at a statistically significant level. The conclusion in regard to dose equivalence was that daily doses of 2 Gy should be replaced by 3 doses of 0.78 Gy given at 4-hours intervals. In this study, the magnitude of the improvement in survival depended on whether or not the patient had undergone surgical resection of the bulk of the tumor. Among patients who had only undergone a biopsy procedure, superfractionation increased the median survival from 11.7 to 35.1 weeks, whereas among patients who had a surgical resection, it increased the median survival from 35.1 to 48.6 weeks.

Final results are not yet available for any prospective randomized (phase III) trials in which normal tissue effects were equivalent and in which superfractionation was used in the treatment of malignant brain tumors. One randomized trial has been reported [30] in which the dosage was reduced rather than increased for the superfractionation arm; it is encouraging that this trial had a null result. Two additional trials have produced preliminary results that favor superfractionation. In one of these trials [27], the same total dose (50 Gy) was used for both conventional and superfractionated radiation therapy. In this trial, the actuarial 1-year survival for a group of 35 patients who received superfractionation was 54%, as compared to 21% for 34 patients who received conventional fractionation [27]. The results of this trial have not yet reached a level of statistical significance.

Preliminary results are also available on a trial that is currently underway at the University of Alberta. This trial has 3 arms, comparing 58 Gy in 30 fractions in 6 weeks of conventional treatment with 2 superfractionation arms in which 1 group receives a radiation sensitizer, misonidazole. The dosage for the 2 superfractionation arms is 61.4 Gy in 69 fractions, given 3 times/day at 4-hour intervals over 4.5 weeks. The actuarial survival for the 37 patients who have received conventional treatment is 24.6 weeks, whereas it is 55 weeks for those who have received superfractionation alone, and for those who have received both superfractionation and misonidazole it is currently 78.4 weeks [*Urtasun,* pers. commun., 1982]. Although these results are preliminary, they are certainly encouraging.

Superfractionation holds considerable promise. The available evidence suggests that superfractionation significantly prolongs life for patients with malignant brain tumors, although it does not at present suggest that a large proportion of patients with glioblastoma multiforme will be cured with superfractionated radiation therapy alone. Cure for these patients may result when superfractionation is combined with other treatment methods, such as chemotherapy, or with the use of radiation sensitizers, such as misonidazole.

Particle Irradiation of Brain Tumors

In addition to multiple daily fractionation, efforts to find better modalities of treatment of gliomas of the brain have also included the combination of irradiation with chemotherapy or radiosensitizing compounds [2, 35], interstitial brain implants with radioactive iodine-125 seeds, and irradiation with particles such as neutrons, pions, and heavy ions. The use of particle radiation therapy in the treatment of brain tumors has the theoretical advantage of an enhanced biologic effect because of the higher rate of transfer of energy from these heavier particles, which results in overcoming the protective effect of hypoxia on tumor cells as well as diminishing the variation in radiosensitivity among cells in different phases of the cell cycle [9–11].

The biologic and physical parameters of particle beams have been investigated in cell, tissue, and tumor systems in order to focus on the properties desirable for radiation therapy [4, 5, 22, 34, 36]. This work has included studies of the nature of sublethal and potentially lethal lesions and the capabilities of repair of such damage. Radiobiologic studies of tissue have included work on the effects of heavy ions on normal mammalian tissues, such as skin, brain, and spinal cord [22, 33].

At the University of California Lawrence Berkeley Laboratory, radiation therapy with helium and heavier particles of atomic numbers between 10 and 15 is being investigated to determine its efficacy as a therapeutic modality in the treatment of human cancers [4, 15]. These particles have physical and biologic advantages for the delivery of radiation of many cancers, such as improved dose localization and consequent reduction of the dosage to normal tissue adjacent to the tumor, diminished OER, depressed enzymatic repair mechanisms, decreased variation of radiosensitivity during the cell division cycle, greater than expected delay in cell division, and decreased protective effects of neighboring cells in organized systems [12, 32, 36, 37].

Neutron Therapy

A variety of investigations has been performed worldwide to evaluate the merits of neutron therapy in the management of gliomas of the brain. These have included both pilot studies and randomized clinical trials. A fairly wide range of therapeutic dose prescriptions has been used, as well as different neutron beams with varying physical and biologic characteristics.

In one of the first such studies, performed from 1973 through 1977 in 37 patients at the University of Washington School of Medicine [26], a 21-MeV deuteron on beryllium cyclotron-produced beam was used to deliver doses of either neutrons at 150 rad/fraction (range 1,550–1,850 rad) or mixed neutron and photon beams of 720–840 neutron rad and 3,300–4,200 photon rad. In this group, the irradiation volume was reduced for the final 10–20% of the dose. 15 patients had grade III lesions and 22 had grade IV tumors. The survival rate for this group of patients was less than that for patients treated with photon irradiation in prior studies. Autopsy data were available on 15 patients; in 14 cases coagulation necrosis replaced the tumor mass and no viable-appearing tumor was observed. Diffuse gliosis and demyelination of the white matter was found in areas of the brain within the neutron-irradiation volume but distant from the tumor. Thus, although tumor destruction was observed, it was not associated with either a longer or better quality of survival.

Between November 1975 and November 1978, 22 patients with grade III and grade IV glioma of the brain were treated with 14 MeV D, T generator-produced fast neutrons at the Netherlands Cancer Institute [3]. 80% of the patients had clinical evidence of local recurrence and a 1-year survival rate of only 32%. A total equivalent dose of 6,500 rad was given that consisted of a dose of 3,000 photon rad over 3½ weeks delivered to the whole brain, followed by a boost with neutrons of 1,160 rad. The boost treatment was delivered to a volume with a margin of 2 cm around the visible tumor, as seen on CT scans and angiograms. In this group of patients, there was no evidence of improvement over the status for the historical controls, however, there were no complications noted. It was thought that their dose was too low to achieve tumor control. In this study, the neutron doses are expressed as the neutron physical dose plus one-third of the gamma contribution – ordinarily between 10 and 20% of the total dose, depending on field size and depth. This assumes a neutron relative biologic effectiveness of 3.

In a randomized study done at the Medical Research Council Cyclotron Unit, Hammersmith Hospital and Royal Marsden Hospital, London [13], 63 patients with glioblastoma received either fast-neutron or X-ray therapy. The neutron therapy was given with neutrons of 7.5 MeV mean energy produced by bombarding a beryllium target with 16 MeV cyclotron-produced deuterons. 18 patients received 1,560 rad in 12 fractions in 26–33 days; 11 patients received 1,300 rad and 1 patient had 1,400 rad given in the same overall time and with the same number of fractions. Partial-brain, large-volume irradiation was used. The patients receiving photon irradiation were treated with 6 MeV X-rays to doses of 5,000–5,500 rad in 25–30 fractions in 5–6 weeks. There was no difference in survival between the 2 groups, although histologic examination showed less tumor at autopsy or second craniotomy in the neutron-treated group than in the photon-treated group. In patients in whom autopsy was performed, there was either no microscopic residual tumor or minimal residual tumor in 11 of 16 neutron-treated patients (69%), whereas such a favorable result was found in only 1 of 7 photon-treated patients; moderate to gross residual tumor was present in 5 of 16 neutron-treated patients and in 6 of 7 photon-treated patients. Among the neutron-treated patients, a smooth-walled, cystic cavity without tumor was found in 2 patients. Vascular changes were present. Of note is that several areas of brain outside the tumor but within the neutron-irradiation volume showed diffuse and focal degenerative changes in the white matter. At all neutron dose levels, there was essentially the same degree of tumor destruction.

In a study performed at the Middle Atlantic Neutron Therapy Association-Naval Research Laboratory 76-inch Isochronous Cyclotron [23], a group of 31 patients with malignant brain tumors was treated with neutrons having an energy of 15 MeV. All but 1 of these patients had a tissue diagnosis and were classified as having either grade III or grade IV glioma. The 30 patients with histologic diagnosis included 17 patients who were treated with neutrons alone and 10 patients who received a planned neutron boost of approximately 600 neutron rad in 6–8 fractions following 5,000 photon rad irradiation to the whole brain in 5 weeks. 3 neutron therapy patients received treatment after 2,000, 3,000 and 4,000 rad of conventional irradiation; in this group, neutron doses varied from 725 to

1,120 rad (neutron plus gamma). Overall, the median survival was 11 months in the patients with grade III tumors and 6 months in those with grade IV tumors. A median survival of 11 months was found in 17 patients treated with neutrons alone.

Persistent disease was confirmed histologically in 7 of 8 patients who had an autopsy. However, there was a difference between the patients in this series and the University of Washington patients, in that a number of patients had both persistent disease and coagulative necrosis. In this study, no significant brain damage outside the tumor region was identified.

Inasmuch as neither whole-brain irradiation with neutrons nor reduced-volume neutron-boost therapy after photon irradiation of the whole brain has resulted in any improvement in survival as yet, current efforts in neutron therapy focus on exploring different techniques to exploit the ability of neutrons to destroy glioma without the serious sequelae of severe damage to the normal brain.

The current Radiation Therapy Oncology Group trial attempts to apply the potentiation that might occur when neutron (or heavy charged-particle) irradiation is given in close proximity to a low LET photon treatment (within 30 min). In effect, the particle dosage causes lethal damage to some tumor cells and preconditions (causes sublethal damage to) other tumor cells, such that the efficacy of the low LET irradiation should be enhanced. Studies by *Ngo* et al. [28] have demonstrated this phenomenon in cultured cells. Presumably, some differential effect between tumor and normal tissues has to exist for this method to be of clinical value. In the present study, several levels of neutron doses are being evaluated in the hope of maximizing tumor destruction while keeping damage to normal brain to an acceptable level.

Such an approach to the treatment of glioblastoma is also under clinical study at the University of Chicago by *Griem* et al. [pers. commun., 1982]. This type of treatment regime is only practical where the particle and low LET irradiation facilities are located reasonably close together or in the same institution, so that the treatments can be given within a short time of each other.

Charged Particle Therapy of Brain Tumors

Efforts are also underway to study the use of charged particles in the therapy of brain tumors. At the Triumph Pion Facility in Vancouver,

British Columbia [*Goodman,* pers. commun., 1982], a pilot study of brain tumor irradiation is underway. Patients receive treatment with 14 fractions of 125 pion rad, each as a boost to conventional photon treatment. The irradiation fields are in the order of 6×6 cm to 8×9 cm.

These patients are treated with a spot raster scanning technique in order to localize the dose generally to the tumor volume plus a margin of normal tissue, which is included to eradicate possible microscopic extension of tumor. Results from this study are not yet complete enough to be evaluated.

Pions have been used for the treatment of grade III and grade IV gliomas of the brain at Los Alamos Meson Physics Facility [8] since 1976. Some patients receive boost therapy after photon irradiation and others receive therapy with pions alone. The results in this small series are of interest, as there is a median survival time of 12 months for the patients with grade IV astrocytoma and of 16 months for those with grade III lesions. Further time is required to evaluate these data fully; perhaps the information from future studies of patients irradiated with pions at the Swiss Institute of Nuclear Research will add to our knowledge.

Helium and heavier particles, such as carbon, neon, silicon and argon ions, in the irradiation of human cancers are being investigated at the University of California Lawrence Berkeley Laboratory in a systematic study of the role of such particles in clinical practice [9–12, 15, 31].

Since 1975, 11 patients have completed irradiation with helium ions for grade III or grade IV primary malignant glioma of the brain. Most of these patients have received boost therapy following photon irradiation. The total dosages have ranged from 60 to 64 Gray equivalents (GyE) in 7–8 weeks given at 2.0 GyE/fraction, 4 fractions/week. The median survival in this group is 16 months; 2 patients, both of whom had grade III lesions, remain free of recurrence at 30 and 34 months posttreatment, respectively.

More recently, a small group of 16 patients has been treated either with carbon or neon ions as a boost to photon therapy or entirely with neon ions. In a group of 8 patients treated solely with neon-ion therapy, the initial dose selected was 48.0 GyE in 16 fractions over 4 weeks. Most of these patients failed rapidly; the median survival time was only 6 months (neon). 5 of 8 patients already have tumor recurrence and the follow-up period in the other patients is short. The tumor dose for these patients may have been too low, and it will be raised gradually to deliver at least 54 GyE in 16 fractions.

Carefully integrated studies with the Brain Tumor Research Center at the University of California, San Francisco, will be done, including serial follow-up quantitative CT scanning to detect effects on both normal brain and tumor [20]. Nuclear magnetic resonance scanning and positron emission scanning also are being performed in collaboration with the Research Medicine Group at the Lawrence Berkeley Laboratories, where research in the use of such scanning techniques has been done by *Ganz* et al. [21]. The initial efforts have been rewarding, showing good correlation of tumor extent with that seen with CT scans and other diagnostic modalities. Future possibilities include particle radiographic scans that may be performed in situ just before the patient receives irradiation with heavy charged particles. Such scans may be performed with autoradioactive beams, such as carbon-13 or neon-19, which would allow the possibility of localizing the high-dose volume to be delivered to the target tumor volume. A positron-emitting beam analyzer has been developed at Lawrence Berkeley Laboratories under the direction of *Chatterjee* et al. [14] that promises to afford the ability of detecting the presence of such beams as neon-19, thus assisting in localizing the spread-out Bragg peak and assuring its coincidence with the desired tumor target. With sufficient development, such a device could be operative while the patient is in the treatment position, thus affording a check of correct treatment parameters during the actual therapy. A sufficiently intense autoradioactive beam, such as neon-19, might even be used for the actual treatment, particularly if fully 3-dimensional beam scanning is developed for clinical use. Such techniques for beam scanning are already under development at Lawrence Berkeley Laboratories with charged particles, and in Japan with proton beams [1].

Conclusions regarding the results obtained with particle therapy must await further study of several factors. Further radiobiologic data on their mechanisms of action are needed, as well as data regarding late effects, the selection of appropriate patients, and tumor types. In addition, phase I and II studies of various particles must be performed, and fractionation schemes and planned improvements in beam delivery for neutrons and heavy ions in hospital-based machines must be developed.

References

1 Akanuma, A.; Morita, S.; Ishikawa, T.; Tsunemoto, H.; Kawachi, K.; Kanai, T.; Hiraoka, T.; Nakamura, Y.; Furukawa, S.: Proton beam radiotherapy at National

Institute of Radiological Science. USA-Japan Cooperative Cancer Research Program Symposium, Chiba, Kyoto 1982.
2 Ang, K.K.; Van der Scheueren, E.; Notter, G.; Horiot, J.C.; Chenal, C.; Fauchon, F.; Raps, J.; Van Peperzeel, H.; Goffin, J.C.; Vessière, M.; Van Glabbeke, M.: Split course multiple daily fractionated radiotherapy schedule combined with misonidazole for the management of grade III and IV gliomas. Int. J. Radiat. Oncol. Biol. Phys. *8:* 1657–1664 (1982).
3 Battermann, J.J.: Fast neutron therapy for advanced brain tumors. Int. J. Radiat. Oncol. Biol. Phys. *6:* 333–335 (1980).
4 Blakely, E.A.; Ngo, F.Q.H.; Chang, P.Y.; Lommel, L.; Tobias, C.A.: Silicon: radiobiological cellular survival and the oxygen effect. Lawrence Berkeley Lab. Rep. LBL-11220, pp. 119–124 (1980).
5 Blakely, E.A.; Tobias, C.A.; Ngo, F.Q.H.; Yang, T.C.H.; Smith, K.C.; Lyman, J.T.: Inactivation of human kidney cells by high energy monoenergetic heavy ion beams. Radiat. Res. *30:* 122–160 (1979).
6 Brown, J.M.: Exploitation of kinetic differences between normal and malignant cells. Radiology *114:* 189–197 (1975).
7 Brown, J.M.: The shape of the dose-response curve for radiation carcinogenesis, extrapolation to low doses. Radiat. Res. *71:* 34–50 (1977).
8 Bush, S.E.; Smith, A.R.: Current status of pion radiotherapy at Lampf; in Skarsgard, Pion and heavy ion radiotherapy: preclinical and clinical studies (Elsevier Biomedical, New York 1983).
9 Castro, J.R.: Particle therapy: the first forty years. Semin. Oncol. *8:* 103–109 (1981).
10 Castro, J.R.; Quivey, J.M.; Lyman, J.T.; Chen, G.T.Y.; Phillips, T.L.; Tobias, C.A.: Radiotherapy with heavy charged particles at Lawrence Berkeley Laboratory. J. Can. Ass. Radiol. *31:* 30–34 (1980).
11 Castro, J.R.; Quivey, J.M.; Lyman, J.T.; Chen, G.T.Y.; Phillips, T.L.; Tobias, C.A.; Alpen, E.L.: Current status of clinical particle radiotherapy at Lawrence Berkeley Laboratory. Cancer *46:* 633–641 (1980).
12 Castro, J.R.; Saunders, W.M.; Tobias, C.A.; Chen, G.T.Y.; Curtis, S.; Lyman, J.T.; Collier, J.M.; Pitluck, S.; Woodruff, K.A.; Blakely, E.A.; Tenforde, T.; Char, D.; Phillips, T.L.; Alpen, E.L.: Treatment of cancer with heavy charged particles. Int. J. Radiat. Oncol. Biol. Phys. *8:* 2191–2198 (1982).
13 Catterall, M.; Bloom, J.; Ash, D.; Walsh, L.; Richardson, A.; Uttley, D.; Gowing, N.; Lewis, P.; Chauser, B.: Fast neutrons compared with megavoltage X-rays in the treatment of patients with supratentorial glioblastoma: a controlled pilot study. Int. J. Radiat. Oncol. Biol. Phys. *6:* 261–266 (1980).
14 Chatterjee, A.; Saunders, W.; Alpen, E.; Alonso, J.; Scherer, J.; Llacer, J.: Physical measurements with high-energy radioactive beams. Radiat. Res. *92:* 230–244 (1982).
15 Chen, G.T.Y.; Castro, J.R.; Quivey, J.M.: Heavy charged particle radiotherapy. Annu. Rev. Biophys. Bioeng. *10:* 419–429 (1981).
16 Douglas, B.G.: Implications of the quadratic cell survival curve and human skin radiation 'tolerance doses' on fractionation and superfractionation dose selection. Int. J. Radiat. Oncol. Biol. Phys. *8:* 1135–1143 (1982).
17 Douglas, B.G.: Superfractionation: its rationale and anticipated benefits. Int. J. Radiat. Oncol. Phys. *8:* 1143–1153 (1982).

18 Douglas, B.G.; Worth, A.J.: Superfractionation in glioblastoma multiforme – results of a phase II study. Int. J. Radiat. Oncol. Biol. Phys. *8:* 1787–1794 (1982).

19 Douglas B.G.; Fowler, J.F.: The effect of multiple small doses of X-rays on skin reactions in the mouse and a basic interpretation. Radiat. Res. *66:* 401–426 (1976).

20 Fike, J.R.; Cann, C.E.; Davis, R.L.; Philips, T.L.: Radiation effects in the canine brain evaluated by quantitative computed tomography. Radiology *144:* 603–608 (1982).

21 Ganz, E.; Budinger, T.; Mathis, C.; Yano, Y.; Huesman, R.: Dynamic positron emission tomography and the analysis of regional metabolism in human gliomas. Annual Meeting of the American Association of Neurological Surgeons, 1982.

22 Goldstein, L.S.; Phillips, T.L.; Fu, K.K.; Ross, G.Y.; Kane, L.J.: Biological effects of accelerated heavy ions. I. Single doses in normal tissues, tumors and cell in vitro. Radiat. Res. *86:* 529–554 (1981).

23 Herskovic, A.H.; Ornitz, R.D.; Shell, M.; Rogers, C.C.: Glioblastoma multiforme treated with 15 MeV fast neutrons. Cancer *49:* 2463–2465 (1982).

24 Howard, A.: The role of oxygen in the repair process; in Bond et al., Conference on Time and Dose Relationships in Radiation Biology as Applied to Radiotherapy, Carmel, Calif., 1969. Brookhaven Natn. Lab. Rep. BNL 50203 (C-578), pp. 70–81 (Brookhaven National Laboratory, Upton 1970).

25 Kelleter, A.M.; Rossi, H.H.: The theory of dual radiation action. Curr. Top. Radiat. Res. *8:* 85–158 (1972).

26 Laramore, G.E.; Griffin, T.W.; Gerdes, A.J.; Parker, R.G.: Fast neutron and mixed (neutron/photon) beam teletherapy for grades III and IV astrocytomas. Cancer *42:* 96–103 (1978).

27 Muller, P.J.; Shin, M.D.: Superfractionation radiotherapy of malignant astrocytomas: preliminary experimental and clinical results. Can. J. Surg. *25:* 415–416 (1982).

28 Ngo, F.Q.H.; Blakely, E.A.; Tobias, C.A.: Sequential exposures of mammalian cells to low and high-LET radiations. I. Lethal effects following X-ray and neon ion irradiation. Radiat. Res. *87:* 59–78 (1981).

29 Palcic, B.; Browsing, J.W.; Skarsgard, L.D.: Survival measurements at low doses; oxygen enhancement ratio. Br. J. Cancer (in press, 1983).

30 Payne, D.G.; Simpson, W.J.; Keen, C.; Platts, M.E.: Malignant glioma: hyperfractionated and standard radiotherapy and chemotherapy in a randomized prospective clinical trial. Int. J. Radiat. Oncol. Biol. Phys. *7:* 1217 (1981).

31 Quivey, J.M.; Castro, J.R.; Chen, G.T.Y.; Moss, A.A.; Marks, W.M.: Computerized tomography in the quantitative assessment of tumor response. Br. J. Cancer *41:* suppl. IV, pp. 31–34 (1980).

32 Raju, M.R.; Bain, E.; Carpenter, S.G.; Cox, R.A.; Robertson, J.B.: A heavy particle comparative study. II. Cell survival versus depth. Br. J. Radiol. *51:* 704–711 (1978).

33 Rodriguez, A.; Alpen, E.L.: Cell survival in spheroids irradiated with heavy ion beams. Radiat. Res. *85:* 24–37 (1981).

34 Roots, R.; Yang, T.; Craise, L.; Blakely, E.; Tobias, C.A.: Rate of rejoining of DNA breaks induced by accelerated carbon and neon ions in the spread Bragg peak. Int. J. Radiat. Res. *38:* 203–210 (1980).

35 Simpson, W.J.; Platts, M.E.: Fractionation study in the treatment of glioblastoma multiforme. Int. J. Radiat. Oncol. Biol. Phys. *1:* 639–644 (1976).

36 Tobias, C.A.; Blakely, E.A.; Alpen, E.L.; Castro, J.R.; Ainsworth, E.J.; Curtis, S.B.; Ngo, F.Q.H.; Rodriguez, A.; Roots, R.J.; Tenforde, T.; Yang, T.C.H.: Molecular and cellular radiobiology of heavy ions. Int. J. Radiat. Oncol. Biol. Phys. *8:* 2109–2120 (1982).
37 Tobias, C.A.; Blakely, E.A.; Ngo, F.Q.H.; Yang, T.C.H.: The repair-misrepair model of cell survival; in Meyn, Withers, Radiation biology in cancer, pp. 195–230 (Raven Press, New York 1980).
38 Withers, H.R.: Cell cycle redistribution as a factor in multifraction irradiation. Radiology *114:* 199–202 (1975).

Prof. J.R. Castro, MD, Department of Radiation Oncology, School of Medicine, University of California, Berkeley, CA 94720 (USA)

Prog. exp. Tumor Res., vol. 28, pp. 166–182 (Karger, Basel 1984)

Stereotactic Interstitial Brachytherapy for Malignant Brain Tumors

P. H. Gutin, M. Bernstein

Section of Functional and Stereotactic Surgery, and the Brain Tumor Research Center of the Department of Neurological Surgery and the Department of Radiation Oncology, University of California School of Medicine, San Francisco, Calif., USA

Introduction

More than 90% of anaplastic astrocytomas and glioblastomas are localized to a single area of the brain [23]. Metastases from these tumors within the central nervous system are uncommon [7], and systemic metastases are rare [1]. Most solid tumors that have not metastasized by the time they are detected can be cured by surgery and radiation therapy [49]. Yet, despite the localized nature of most malignant brain tumors, the principal experimental thrust in their treatment has been systemic chemotherapy, a modality for which the traditional target has been metastatic disease [14].

Because a primary brain tumor is most often a localized disease, hyperthermia, intratumoral chemotherapy, and interstitial irradiation, the stereotactic placement of radioactive sources directly into brain tumors, have been suggested as possible local treatments for brain tumors. Interstitial irradiation seems a promising therapeutic approach, given the known efficacy of radiation therapy against brain tumors.

The Brain Tumor Study Group (BTSG) has shown that patients harboring primary malignant brain tumors who received more than 5,000 rad of radiation to the whole brain survived 20.5 weeks longer than patients treated by surgery only [47]. *Walker* et al. [48] analyzed the BTSG data more critically and showed stepwise increments in survival in patient cohorts receiving 5,000, 5,500, or 6,000 rad. It is well known that doses of photon radiation greater than these put the patient at significant risk for developing brain necrosis [42]. Because radiation therapy has been the most effective modality against malignant brain tumors and because necrosis of normal brain has prevented delivery of more than 6,000–7,000 rad at conventional dose rates, brachytherapy becomes a logical local treatment for brain tumors.

Rationale for Interstitial Irradiation

'Conventional' radiation from an X-ray or cobalt source is delivered at dose rates of approximately 200 rad/min; in standard radiation therapy regimens, doses are fractionated and the total course of therapy is given over several weeks or months. In most interstitial irradiation regimens used clinically, peripheral minimum tumor doses are delivered at much lower dose rates, often on the order of 1 rad/min, and the desired dose can be delivered in a few days to a week [41]. Because of its increased therapeutic index relative to external beam treatments, therapeutic doses can be delivered in a much shorter time in brachytherapy regimens. This advantage accrues because of the favorable radiobiology of low dose rate irradiation, and the favorable radiophysics of intratumoral placement of the radioactive source [3].

Radiobiology of Low Dose Rate Irradiation

The dose rate is a major determinant of the biological effect of a given dose of ionizing radiation [18]. In general, as the dose rate is lowered (and the exposure time is prolonged), the biological effect of a given dose of radiation is reduced. This effect is reflected in the radiation survival curve. As the dose rate is lowered, the slope of the survival curve is progressively reduced and the extrapolation number approaches unity. As the dose rate is lowered to approximately 1 rad/min, there is little additional dose rate effect. However, if the dose rate is reduced to below 1 rad/min, a point is reached where mitosis is no longer inhibited (the critical dose rate). This varies from cell system to cell system and depends on the cell cycle time. It has been shown in vivo and in vitro that the amount of radiation absorbed per cell cycle is the critical factor affecting inhibition of mitosis; 720–990 rad/cell cycle inhibits mitosis in most systems studied [19].

The dose rate effect has been observed in most in vivo and in vitro systems and is most pronounced for dose rate of 1–100 rad/min. The dose rate effect is principally the result of the ongoing repair of sublethal radiation damage that occurs during low but not high dose rate exposures [3], which accounts for the progressively reduced dose rates that yield reduced slopes on radiation survival curves.

The principal reason that low dose rate exposure relatively spares normal tissue is that normal tissue appears to repair sublethal damage more efficiently than neoplastic tissue does [11, 24, 25]. The dose rate effect is therefore more pronounced for normal tissue, and more tumor cells will be killed than normal cells for a given radiation dose delivered at low dose rates.

Other factors can accentuate the relative sparing of normal tissue during brachytherapy. For example, most tumors (and not normal tissues) seem to have significant fractions of hypoxic cells [17]. Tumors exposed to low dose rate radiation reoxygenate with time, which causes tumor cells to become progressively more radiosensitive [24].

Intratumoral Placement of Source

The advantages of interstitial therapy over external radiation are apparent. Because of the inverse-square law and tissue attenuation, both of which cause a rapid fall-off of radiation outside the target volume, a radioactive source placed directly into a tumor can deliver a maximum therapeutic dose of radiation to the tumor with relative sparing of the surrounding normal tissues. The inverse-square law states that, for any point source of radiant energy, the dose absorbed is inversely proportional to the square of the distance away from the source. Tissue attenuation is determined by the density of the tissue and the energy of the radiation; the less energetic the radiation, the less tissue it will penetrate and vice versa.

History of Brain Tumor Brachytherapy

The potential uses of brachytherapy for the treatment of intracranial neoplasia were recognized early in this century. In 1912, after a trans-sphenoidal operation on a young woman with acromegaly, *Hirsch* [22] inserted a radium-loaded probe through the nasal cavity into the sella turcica; even though this patient did not improve, he continued to use the technique, and for over 4 decades he had great success treating pituitary adenomas and craniopharyngiomas with irradiation from radium. Radioactive phosphorus (^{32}P) and then ^{90}Y eventually replaced radium for the treatment of craniopharyngioma.

In 1914, *Frazier* [8] first implanted radiation sources into cerebral parenchymal tumors using 50–100 mg of radium implanted for up to 20 h. This was usually combined with external irradiation. Because results were encouraging, by 1922, *Frazier* had used radium to treat 32 patients harboring brain tumors. During the next 4 decades, many series were reported, but few addressed in depth the methodology and efficacy of the procedure.

It was not until the introduction of the *Leksell* [31] stereotactic system in the late 1940s that clinical stereotactic surgery became popular and made possible the accurate injection of ^{32}P into cystic craniopharyngiomas [2]. More importantly, stereotactic techniques made possible the placement of radioactive sources into deep, inoperable tumors such as thalamic gliomas [44]. *Talairach* (Paris) and *Mundinger* (Freiburg) are the pioneers of stereotactic interstitial brachytherapy, both of whom have extensive experience with the use of brachytherapy for many types of brain tumors [38, 46]. In general they have treated gliomas, which were often low-grade tumors, with relatively low doses delivered from permanently implanted, relatively low activity sources that in recent years were primarily ^{192}Ir.

The computerized tomography (CT) scanner has revolutionized the use of interstitial brachytherapy for brain tumors. Before the availability of CT, tumors were localized and doses were planned on the basis of films obtained by pneumoencephalography, angiography, and nuclear scanning. In addition to providing a more exact preoperative estimation of the size, consistency, and location of the tumor, the CT scan has made dosimetric calculations much more exact, and the use of the CT scanner with stereotactic head frames has made the accurate placement of radioactive sources into tumors more reliable and facile. The use of CT-aided stereotactic cerebral brachytherapy for brain tumors has been described [37, 39].

Interstitial Brachytherapy for Malignant Gliomas

We believe that the initial treatment for low-grade gliomas should be histological verification followed by standard radiation therapy [30]. Brachytherapy has no place in the initial treatment of these tumors but may be helpful in carefully selected recurrences [15, 16]. Likewise, we believe that brain metastases from systemic cancer should be irradiated initially by external techniques; brachytherapy is of value only for a metastasis that recurs in 1 cerebral location several times, and if there is no evidence of systemic disease.

Isotopes

A number of isotopes, primarily ^{198}Au and ^{192}Ir, have been used for brain tumor brachytherapy [3, 43], but ^{125}I, which has only recently been used for brachytherapy for disease at several sites, particularly carcinoma of the prostate [27, 45], seems to be the best isotope for brain tumor brachytherapy. ^{125}I emits characteristic X-rays with energies of 27–35 keV, which are much lower than the energy ranges of radiation from other isotopes such as ^{192}Ir and ^{198}Au that are commonly used for brachytherapy [29]. Because of its low energy, the X-radiation from implanted ^{125}I can be attenuated effectively by interposed tissues [29], which spares normal tissues during ^{125}I brachytherapy. In addition, less radiation exits the skull, and X-rays that do escape are more effectively shielded [36]. When ^{125}I is used for brachytherapy, surgeons, nurses, and other personnel can protect themselves with lead aprons, a maneuver that will not protect personnel from the much higher energy radiation emitted by other isotopes used for brachytherapy. Another favorable quality of ^{125}I, possibly the result of the low energy emissions, is its purportedly higher relative biological effectiveness compared to isotopes such as ^{192}Ir [9, 27].

^{125}I is provided as a standard low activity (0.5 mCi) source which, if implanted in reasonable numbers into fast-growing, malignant brain tumors, delivers radiation at dose rates that are insufficient to deter the growth of these tumors [15, 16]. Higher activity ^{125}I sources (approximately 40 mCi) are provided only by special order and at considerable expense. Other disadvantages arising from the use of ^{125}I are attributable to the same characteristic which accounts for its many advantages, low energy. It is difficult to calibrate the individual sources because of their low energy emissions, and the welded ends on the small tubular sources are sufficiently thick that they shield radiation along the source axis and give rise to anisotropy, which complicates the determination of exact dosimetry [29].

Patient Selection

Patients with malignant gliomas selected for brachytherapy should not be neurologically devastated. Patients who are confused, hemiplegic, and/or incontinent require extensive, close-contact nursing care; nurses caring for these patients while sources are implanted would be exposed to unacceptable levels of radiation. Therefore, patients with these clinical conditions are not selected for treatment with brachytherapy.

Tumors selected for treatment should be localized, preferably with margins that are well-defined on the contrast-enhanced CT scan. There is no upper limit on tumor size for brachytherapy because careful treatment planning allows reasonable radiation coverage of even very large lesions with the addition of more sources spaced throughout the tumor. However, in our experience, lesions larger than 6 cm in their greatest dimension generally are not localized and well-defined. Tumors in sensitive locations of the brain should be treated cautiously and conservatively, but often there are no reasonable alternative treatments. We have little experience with tumors in the brain stem or cerebellum.

Removable Implants

For successful treatment of rapidly expanding (malignant) brain tumors, the dose rate of radiation delivered must be sufficiently high that tumor cells receive a critical threshold dose during a complete cell cycle [21]. This condition can be met by using as radioactive sources high-activity radioisotopes that can be removed from the brain after the dose is delivered; if the sources remain implanted, normal brain surrounding the source will be exposed to potentially toxic doses of radiation. Radioactive sources that are encased in catheters can be removed after the desired dose is delivered; moreover, catheters can hold the source with precision at the tumor target site in the often semisolid or liquid centers of more grossly necrotic malignant brain tumors. Our coaxial catheter system, which provides for verification of catheter position before afterloading the radioactive sources, has been described [13].

Stereotactic Techniques

An integrated CT-stereotactic system is absolutely essential for the interstitial brachytherapy of brain tumors [37]. The integration of stereotactic systems with CT scanning technology has allowed pinpoint accuracy in the biopsy, or implantation of, isotopes into intracranial tumors [37, 39, 40] (fig. 1). The availability of such systems has made implantation of isotopes into tumors at open craniotomy suboptimal from several standpoints. Most importantly, it is frequently difficult for the surgeon to adequately expose a glioma at craniotomy to be able to appreciate its geometry. This makes the implantation of a suitable isotope nearly impossible. In addition, it may take several hours to close a

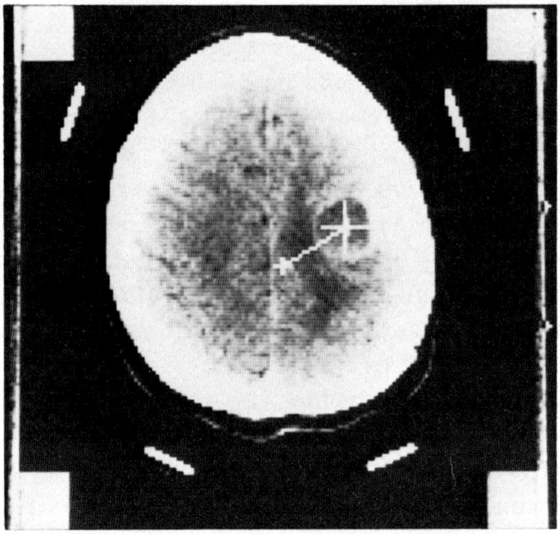

1a

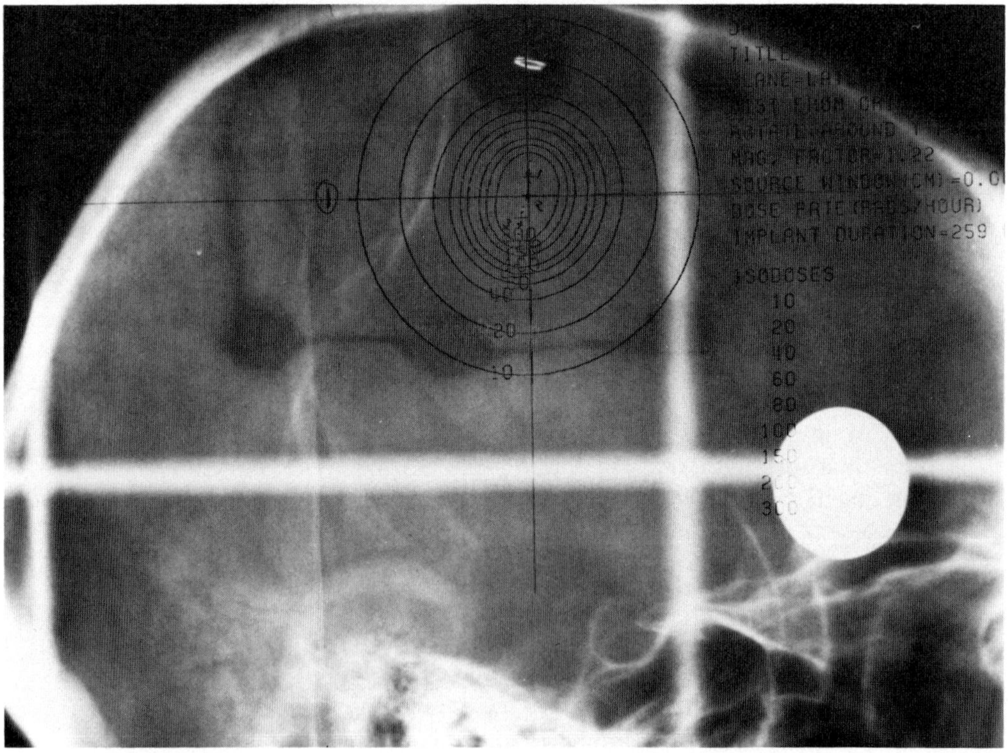

1b

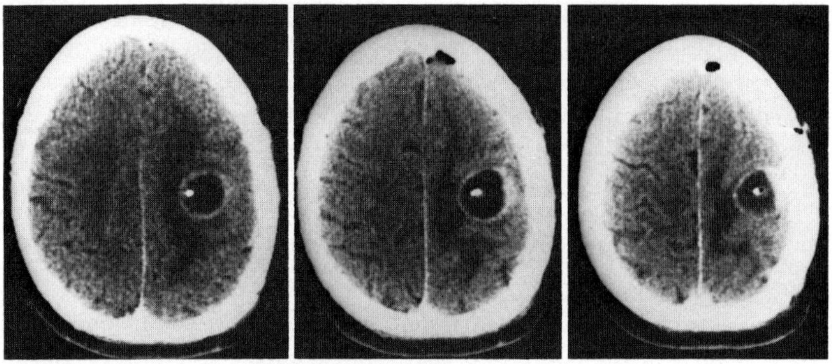

1c

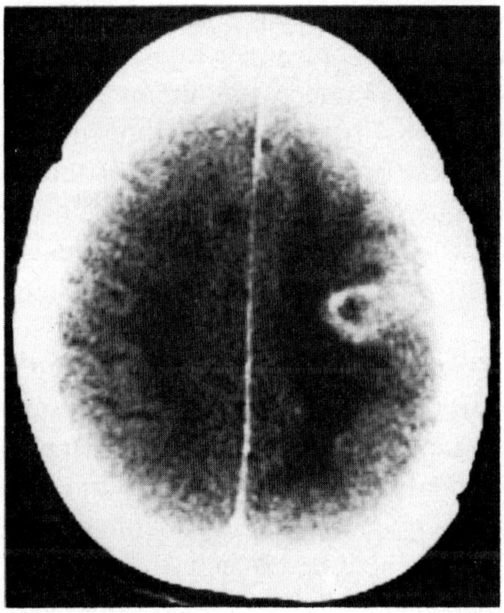

1d

Fig. 1. CT scans and a radiograph for a patient harboring a left parietal glioma. *a* CT scan taken with the plastic Leksell stereotactic frame in position. The four vertical (square) posts of the frame are visible. The geometric center of the frame (small cross) is related to the target for ^{125}I placement by the scanner's computer, and coordinates for the target are generated [37]. *b* Plain radiograph showing three ^{125}I sources implanted in a single catheter into the tumor; isodose curves for this combination of sources are superimposed on the radiograph. *c* CT scans of progressively superior portions of the brain that show the positions of each of the sources implanted transversely through the center of the tumor. *d* CT scan taken 2 months after implantation that shows the reduction in the size of the tumor.

craniotomy wound, during which time the radiation exposure to surgeons and nurses would be greater than necessary. By contrast, isotope implantation using stereotactic techniques can be done through small skin incisions and burr or twist drill holes, which makes closure of the wounds facile.

A number of integrated CT-stereotactic systems have been described [37, 39], all of which seem satisfactory. We have adapted the Leksell system because of its simplicity and the great freedom it allows in selecting an approach to tumor targets [37].

We prefer to do the actual placement of the catheters and loading of the isotope in the controlled environment of the operating room rather than in the CT scanner room. After the frame is applied and localizing scans and calculation of stereotactic coordinates are completed, the patient is taken to the operating room where under local anesthesia the catheters are placed through burr holes, loaded with dummy sources, and localized on intraoperative radiographs. If they are positioned correctly compared to reformatted sagittal and coronal CT images of the tumor, they are loaded with the coaxial inner catheters housing the ^{125}I sources. Surgeons and nurses wear lead aprons and radiation exposure badges for this part of the procedure. Protective leaded glasses are advisable for the surgeon.

After catheters have been loaded, surgical wounds are closed to totally internalize the system and the patient is isolated in a private room and cared for by nurses trained in radiation safety. The radiation physicist monitors the room carefully and dictates maximum exposure times for the nurses. Nurses wear lead aprons when tending to the patient and/or stand behind a lead shield placed at the bedside. When the dosimetry is finalized, and the desired dose has been delivered (generally 7–10 days), the patient is taken to the operating room and, under local anesthesia, the catheters are removed.

Dosimetry

On the day after the implant procedure, the patient has orthogonal radiographs of the skull taken for source localization. Isodose curves are drawn when the source strengths and locations are entered into a computer program for ^{125}I dosimetry [15] (fig. 1b). A CT scan is performed to verify accurate positioning of sources (fig. 1c).

Because the dosimetry is generated from source positions determined on plain radiographs, the exact relationship between the isodose curves and the tumor margins is unclear. Isodose curves can be reduced to the magnification of the CT scan and superimposed on the tumor, but this is at best an approximation. We are designing a computer program that will generate dosimetry from source positions localized on CT scan such that the isodose curves can then be directly superimposed on the scan image of the tumor. Such a program could be used for preoperative planning of optimal implant locations and to determine final dosimetry based on source visualization on a postoperative scan.

Minimum (peripheral) tumor doses delivered have ranged from 3,000 to 12,000 rad in our series. Most recently we have given doses of approximately 10,000 rad.

Evaluation of Patients

Corticosteroid doses are adjusted as needed to improve neurologic function and to reduce the symptoms of increased intracranial pressure. Because improvement caused by steroids can mimic response to interstitial radiation, doses are increased only when required to treat clear-cut clinical deterioration. In addition, attempts are made to reduce the steroid dose every 6–8 weeks if the patient is clinically stable or improving.

Patients are evaluated by neurological examinations and CT scans at intervals of 8 weeks, when possible, and graded on a scale of −2 to +2, which ranks deterioration or improvement, respectively [32]. Response is defined as a definitive improvement in at least 1 of the 2 criteria in the same evaluation period with at least stability in the other if the corticosteroid dose is stable or decreasing. Progression of disease is defined as definite deterioration in at least 1 of the 2 criteria with steroid doses stable or increasing. Stable disease is defined as no definitive change in either test with an unchanged or decreased steroid dose. Time to progression is measured from the day of implantation until progression (as defined above) is determined.

Treatment of Recurrent Malignant Gliomas

These criteria for response are similar to those that have been used to evaluate patients receiving chemotherapy for brain tumors [32]. These criteria have proved difficult to apply to patients receiving brachytherapy. It is well known that the CT scan changes and clinical picture of radiation necrosis can be indistinguishable from those of brain tumor progression [5, 12, 28]. It has been shown by others that radioactive implants into gliomas can result in the delayed onset of focally increased contrast

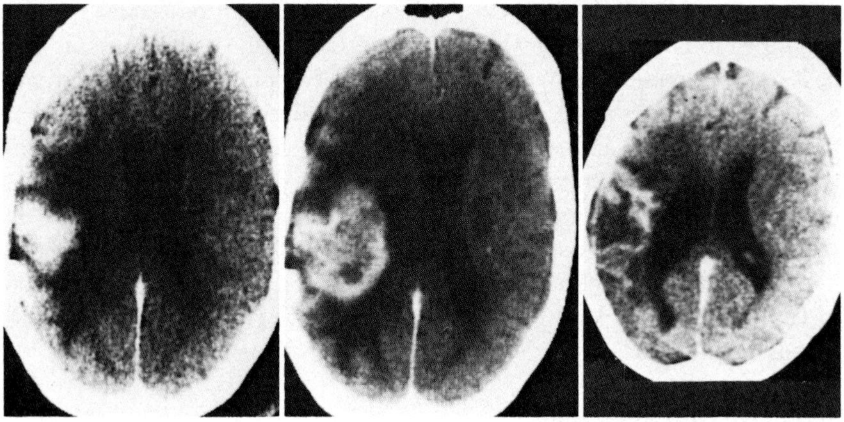

a, b, c

Fig. 2. CT scans of a patient with a recurrent right parietal glioblastoma taken before *(a)* and 5 months after *(b)* interstitial irradiation with high activity ^{125}I (11,000 rad). The lesion, resected 5 months after implantation, was predominantly radiation necrosis. A small amount of tumor was identified in the histological specimen, but there is no evidence of recurrence 7 months after resection (1 year after implant) *(c)*.

enhancement and cerebral edema on CT scan with concomitant neurological deterioration. Stereotactic biopsies have confirmed the diagnosis of radiation necrosis [43]. Differentiating brain tumor progression from focal radiation necrosis on radiographic or clinical findings alone makes response criteria difficult to formulate for patients undergoing brachytherapy.

In practice, more than half of treated patients with recurrent malignant gliomas will show a clear-cut response, as defined by our criteria, with durations of 4 months to more than a year (fig. 1d). Another 15% of patients will stabilize for up to a year, and the remainder will show progression of disease as defined by our criteria, some developing disease in distant areas of the brain and others developing focal radiation necrosis, the apparent explanation for the early CT scan and neurological deterioration described above. Several of these latter patients have been reoperated at 6–12 months after implant, and histological changes consistent with radiation necrosis have been identified, sometimes mixed with apparently viable tumor (fig. 2). Two young men with glioblastomas are now surviving free of disease at 1 year after reoperation, 2 years after implant, and 3 and 4 years after initial diagnosis.

Nearly half of the more than 40 patients with recurrent malignant brain tumors treated with removable ^{125}I implants over the past 3 years are still surviving. Because surviving patients are censored from consideration in survival plots [26], such plots are currently meaningless for our series. One measure of the ^{125}I implants is the median follow-up time from implantation. For our glioblastoma group this is 9 months, for anaplastic astrocytomas 10 months, and for the malignant glioma group as a whole 9 months. These numbers are not overall survival, which in all instances is much longer, but only the follow-up time after implantation for recurrence. Because nearly half of these patients are surviving at the time of this report, the results of this treated group will progressively improve. In addition, refinements in patients selection and in implantation techniques and our increasingly aggressive approach to treatment will improve the results.

The inability to apply standard neurooncological endpoints to patients undergoing brain tumor brachytherapy not only makes them difficult to evaluate but also makes comparisons of the results of this technique with those of chemotherapy difficult. Because of focal radiation necrosis, standard calculations such as response rate and time from response to progression that are used to determine the efficacy of brain tumor chemotherapy become less meaningful when they are used to evaluate the results of brain tumor brachytherapy. Some of our patients, who evidenced disease progression by our criteria and were found to have radiation necrosis at reoperation, have survived longer than patients who showed a response by our criteria. Certainly, if comparisons can be made, the 68% response rate (responses and stabilizations) in our current series is comparable to that of the best chemotherapy regimens for recurrent malignant glioma [33–35] and the median follow-up of 9 months for our small group of patients with recurrent glioblastoma – nearly half of whom still survive – suggests that brachytherapy is significantly better than chemotherapy for this tumor [33–35]. Furthermore, because nearly half are still surviving, the median follow-up is constantly increasing. A direct comparison of brain tumor brachytherapy with chemotherapy, however, is not of critical importance. Brain tumor brachytherapy has been tested at our institution as a phase-II oncology trial and the results stand well by themselves. Particularly intriguing are the several patients with localized recurrences treated with brachytherapy who at subsequent reoperations for focal necrosis had no histologically detectable tumor. Certainly, the 2 patients with recurrent glioblastomas who are surviving without active disease 1 year after reoperation and 2 years after implantation offer some hope that cure of localized recurrences of some glioblastomas might be possible. This is especially true if focal radiation necrosis is a surgically remediable problem. After resection of the focal necrotic mass, some patients have excellent neurological improvement while being tapered completely off steroids; these hopeful findings, which are consistent with the limited reports on this problem [6], are encouraging. The possibility of reoperation should be anticipated in every patient.

Future of Brachytherapy for Malignant Gliomas

Cure of any recurrent malignancy is an enormous problem; it is of course better to control the disease initially. In an attempt to gain immediate local control of what is predominately a localized disease [23], we have begun a protocol in the Northern California Oncology Group in which patients undergo brachytherapy with removable ^{125}I 'boosts' immediately after conventional external beam irradiation. *Walker* et al. [48] analyzed the data of the BTSG and showed stepwise increments in survival in patient cohorts receiving 5,000, 5,500, or 6,000 rad of external beam irradiation after surgery; in all patients the disease eventually recurred and they died. It is known that higher doses of external beam irradiation, which might theoretically lead to further increments of survival, carry an unacceptable risk producing of large volume radiation necrosis [42]. Our implant 'boost' after external beam treatment is designed to increase the radiation dose to the tumor while sparing surrounding normal brain. Such a program has the potential to induce long-term remissions or even produce cures of localized gliomas. The complexities of patient evaluation will be obviated in this study because overall survival and perhaps even the cure rate will be the important statistics. The risk of focal radiation necrosis in patients receiving 'boosts' is unknown, but our projection is that necrosis will be surgically manageable. We cannot now predict the extent of neurologic deficit that will be induced by such aggressive therapy.

It is clear from our work with recurrent tumors that ways should be sought to decrease radiation-induced injury to normal brain. A reduction in the interstitial radiation dose with equal or enhanced capability for local tumor control might be feasible if the interstitial therapy is delivered along with radiation sensitizers or with tumor hyperthermia.

Because the oxygen enhancement ratio for radiation delivered at low dose rates is lower than that for acute exposure and because reoxygenation of tumors can occur during protracted (low dose rate) exposures, hypoxic cell radiosensitizers (nitroimidazoles) should produce little or no sensitization during brachytherapy [20, 24]. However, *Fu* et al. [10] found a dose modification in the EMT-6 tumor system treated with misonidazole and irradiated at low dose rates with an external ^{137}Cs source, and we have shown that desmethylmisonidazole produced a similar dose enhancement in the RIF-1 tumor implanted with removable ^{125}I sources [4]. Nitroimidazoles have been tested in only a very limited way during clinical brachytherapy trials and deserve further trials. Another radiosensitizer with a different mode of action (bromodeoxyuridine) is being tested in our animal implant model.

Hyperthermia for brain tumors is considered in detail later in this volume. It should be emphasized that this modality amplifies the effect of radiation dramatically, especially radiation delivered at low dose rates. Clinical trials of hyperthermia used in conjunction with interstitial brachytherapy for systemic cancer have shown encouraging preliminary results.

Acknowledgments

Supported in part by NIH Program Project Grant CA-13525, NIH New Investigator Research Award CA-30024, American Cancer Society Junior Faculty Clinical Fellowship (ACS JFCF-557) and the Sariano Cancer Research Fund.

We thank *Neil Buckley* for editorial assistance.

References

1 Alvord, E.C.: Why do gliomas not metastasize? Archs Neurol. *33:* 73–75 (1976).

2 Backlund, E.-O.: Stereotaxic treatment of craniopharyngiomas. Acta neurochir. *21:* suppl., pp. 177–183 (1974).

3 Bernstein, M.; Gutin, P.H.: Interstitial irradiation of brain tumors: a review. Neurosurgery *9:* 741–750 (1981).

4 Bernstein, M.; Gutin, P.H.; Deen, D.F.; Weaver, K.A.; Levin, V.A.; Barcellos, M.H.: Radiosensitization of the RIF-1 murine flank tumor by desmethylmisonidazole (Ro-05-9963) during interstitial brachytherapy. Int. J. Radiat. Oncol. Biol. Phys. *8:* 487–490 (1982).

5 Crafts, D.; Wilson, C.: Differential diagnosis of tumor regrowth. Semin. Oncol. *2:* 15–17 (1975).

6 Edwards, M. S.; Wilson, C. B.: Treatment of radiation necrosis; in Gibert, Kagan, Radiation damage to the nervous system, pp. 129–143 (Raven Press, New York 1980).

7 Erlich, S. S.; Davis, R. L.: Spinal subarachnoid metastasis from primary intracranial glioblastoma multiforme. Cancer *42:* 2854–2864 (1978).

8 Frazier, C. H.: The effects of radium emanations upon brain tumors. Surgery Gynec. Obstet. *31:* 236–239 (1920).

9 Freeman, M. L.; Goldhagen, P.; Sierra, E.; Hall, E. J.: Studies with encapsulated ^{125}I sources. II. Determination of the relative biological effectiveness using cultured mammalian cells. Int. J. Radiat. Oncol. Biol. Phys. *8:* 1355–1361 (1982).

10 Fu, K. K.; Hurst, S.; Begg, A. C.; Brown, J. M.: The effects of misonidazole during continuous low dose rate irradiation. Cancer Clin. Trials *3:* 257–265 (1980).

11 Fu, K. K.; Phillips, T. L.; Kane, L. J.; Smith, V.: Tumor and normal tissue response to irradiation in vivo: variation with decreasing dose rates. Radiology *114:* 709–716 (1972).

12 Graeb, D. A.; Steinbok, P.; Robertson, W. D.: Transient early computed tomographic changes mimicking tumor progression after brain tumor irradiation. Radiology *144:* 813–817 (1982).

13 Gutin, P. H.; Dormandy, R. H.: A coaxial catheter system for afterloading radioactive sources for the interstitial irradiation of brain tumors. J. Neurosurg. *56:* 734–735 (1982).

14 Gutin, P. H.; Levin, V. A.: Surgery, radiation, and chemotherapy in the treatment of malignant brain tumors; in Thompson, Green, Controversies in neurology, pp. 67–86 (Raven Press, New York 1983).

15 Gutin, P. H.; Phillips, T. L.; Hosobuchi, Y.; Wara, W. M.; MacKay, A. R.; Weaver, K. A.; Lamb, S.; Hurst, S.: Permanent and removable implants for the brachytherapy of brain tumors. Int. J. Radiat. Oncol. Biol. Phys. *7:* 1371–1381 (1981).

16 Gutin, P. H.; Phillips, T. L.; Hosobuchi, Y.; Wara, W. M.; MacKay, A. R.; Weaver, K. A.; Lamb, S.; Hurst, S.: Local treatment of malignant brain tumors by removable stereotactically implanted radioactive isotopes; in Kärcher, Kogelnik, Reinartz, Progress in radio-oncology, vol. 2, pp. 363–369 (Raven Press, New York 1982).

17 Gutin, P. H.; Wara, W. M.; Phillips, T. L.; Wilson, C. B.: Hypoxic cell radiosensitizers in the treatment of malignant brain tumors. Neurosurgery *6:* 565–576.

18 Hall, E. J.: Radiation dose-rate: a factor of importance in radiobiology and radiotherapy. Br. J. Radiol. *45:* 81–97 (1972).

19 Hall, E. J.: Radiobiology for the radiologist; 2nd ed. (Harper & Row, Hagerstown 1978).

20 Hall, E. J.; Lam, Y.-M.: The renaissance in low dose-rate interstitial implants. Front. Radiat. Ther. Oncol. vol. 12, pp. 21–34 (Karger, Basel 1978).

21 Hall, E. J.; Oliver, R.; Shepstone, B. J.; Bedford, J. W.: On the population kinetics of the root meristem of *Vicia faba* exposed to continuous irradiation. Radiat. Res. *27:* 597–603 (1966).

22 Hirsch, O.: Die operative Behandlung von Hypophysistumoren: Nach endonasalen Methoden. Arch. Lar. Rhinol. *26:* 529–686 (1912).

23 Hochberg, F. H.; Pruitt, A.: Assumptions in the radiotherapy of glioblastoma. Neurology *30:* 907–911 (1980).

24 Kal, H. B.; Barendsen, G. W.: Effects of continuous irradiation at low dose-rates on a rat rhabdomyosarcoma. Br. J. Radiol. *45:* 279–283 (1972).

25 Kal, H.B.; Sissingh, H.A.: Effectiveness of continous low dose-rate gamma-irradiation on rat skin. Br. J. Radiol. *47:* 673–678 (1974).

26 Kaplan, E.L.; Meier, P.: Nonparametric estimation from incomplete observations. J. Am. statist. Ass. *53:* 457–481 (1958).

27 Kim, J.H.; Hilaris, B.S.: Iodine-125 source in interstitial tumor therapy. Am. J. Roentg. *123:* 163–169 (1975).

28 Kingsley, D.P.E.; Kendall, B.E.: CT of the adverse effects of therapeutic radiation of the central nervous system. Am. J. Neuroradiol. *2:* 453–460 (1981).

29 Krishnaswamy, V.: Dose distribution around a ^{125}I seed source in tissue. Radiology *126:* 489–491 (1978).

30 Leibel, S.A.; Sheline, G.E.; Wara, W.M.; Boldrey, E.B.; Nielsen, S.L.: The role of radiation therapy in the treatment of astrocytomas. Cancer *35:* 1551–1557 (1975).

31 Leksell, L.: A stereotaxic apparatus for intracerebral surgery. Acta chir. scand. *99:* 299–233 (1949).

32 Levin, V.A.; Crafts, D.; Norman, D.; Hofer, P.B.; Spire, J.-P.; Wilson, C.B.: Criteria for evaluating malignant brain tumor patients undergoing chemotherapy. J. Neurosurg. *47:* 329–335 (1977).

33 Levin, V.A.; Edwards, M.S.; Wright, D.C.; Seager, M.L.; Schimberg, T.P.; Townsend, J.J.; Wilson, C.B.: Modified procarbazine, CCNU, and vincristine (PCV 3) combination chemotherapy in the treatment of malignant brain tumors. Cancer Treat. Rep. *64:* 237–241 (1980).

34 Levin, V.A.; Gutin, P.H.; Wilson, C.B.: Brain tumors; in Greenspan, Clinical interpretation and practice of cancer chemotherapy, pp. 393–408 (Raven Press, New York 1982).

35 Levin, V.A.; Hoffman, W.F.; Pischer, T.L.; Seager, M.L.; Boldrey, E.B.; Wilson, C.B.: BCNU-5-fluorouracil combination therapy for recurrent malignant brain tumors. Cancer Treat. Rep. *62:* 2071–2076 (1978).

36 Liu, J.; Edwards, F.M.: Radiation exposure to medical personnel during iodine-125 seed implantation of the prostate. Radiology *132:* 748–749 (1979).

37 MacKay, A.M.; Gutin, P.H.; Hosobuchi, Y.; Norman, D.: Computed tomography-directed stereotaxy for biopsy and interstitial irradiation of brain tumors: technical note. Neurosurgery *11:* 38–42 (1982).

38 Mundinger, F.: The treatment of brain tumors with interstitially applied radioactive iostopes; in Wang, Paoletti, Radionuclide applications in neurology and neurosurgery, pp. 199–265 (Thomas, Springfield 1970).

39 Mundinger, F.; Birg, W.; Ostertag, C.B.: Treatment of small cerebral gliomas with CT-aided stereotaxic curietherapy. Neuroradiology *16:* 564–567 (1978).

40 Ostertag, C.B.; Mennel, H.D.; Kiesling, M.: Stereotactic biopsy of brain tumors. Surg. Neurol. *14:* 275–283 (1980).

41 Pierquin, B.: The destiny of brachytherapy in oncology. Am. J. Roentg. *127:* 495–499 (1976).

42 Sheline, G.E.; Wara, W.M.; Smith, V.: Therapeutic irradiation and brain injury. Int. J. Radiat. Oncol. Biol. Phys. *6:* 1215–1228 (1980).

43 Szikla, G.: Stereotactic cerebral irradiation. Proc. Inserm Symp. on Stereotactic Irradiations, Paris 1979 (Elsevier/North-Holland Biomedical Press, Amsterdam 1979).

44 Szikla, G.; Peragut, J.C.: Irradiation interstitielle des gliomes. Neurochirurgie *21:* suppl. 2, pp. 187–228 (1975).

45 Sogani, P.C.; De Cosse, J.J.; Montie, J.; Whitmore, W.F.; Grabstald, H.; Hilaris, B.S.: Carcinoma of the prostate: treatment with pelvic lymphadenoectomy and iodine-125 implants. Clin Bull. *9:* 24–31 (1979).
46 Talairach, J.; Ruggiero, G.; Aboulker, J.; David, M.: A new method of treatment of inoperable brain tumours by stereotaxic implantation of radioactive gold – a preliminary report. Br. J. Radiol. *28:* 62–74 (1955).
47 Walker, M.D.; Alexander, E., Jr.; Hunt, W.E.; MacCarty, C.S.; Mahaley, M.S.; Mealey, J.; Norrell, H.A.; Owens, G.; Ransohoff, J.; Wilson, C.B.; Gehan, E.A.; Strike, T.A.: Evaluation of BCNU and/or radiotherapy in the treatment of anaplastic gliomas. A cooperative clinical trial. J. Neurosurg. *49:* 333–343 (1978).
48 Walker, M.D.; Strike, T.A.; Sheline, G.E.: An analysis of dose-effect relationship in the radiotherapy of malignant gliomas. Int. J. Radiat. Oncol. Biol. Phys. *5:* 1733–1740 (1979).
49 Zubrod, C.G.: Chemical control of cancer. Proc. natn. Acad. Sci. USA *69:* 1042–1047 (1972).

P.H. Gutin, MD, Section of Functional and Stereotactic Surgery, Brain Tumor Research Center, Departments of Neurological Surgery and Radiation Oncology, University of California School of Medicine, San Francisco, CA 94143 (USA)

Prog. exp. Tumor Res., vol. 28, pp. 183–197 (Karger, Basel 1984)

Combined Effects of Drugs and Radiation against Tumor Cells

D. F. Deen, P. J. Tofilon

Brain Tumor Research Center of the Departments of Neurological Surgery and Radiation Oncology, School of Medicine, University of California, San Francisco, Calif., USA

Introduction

Because radiation therapy and chemotherapy are proven effective therapies for a variety of tumors, these modalities have been combined in attempts to achieve a greater tumor response than provided by either treatment alone. Therapeutic effectiveness could be improved if tumor cell kill could be increased without increasing normal tissue toxicity.

The ways in which drugs and radiation affect cell kill are complicated and mechanisms by which individual agents kill cells must be known before the mechanisms by which combination treatment kills cells can be determined. While the events that lead to radiation-induced cell kill are fairly well understood, the mechanisms of drug-induced cytotoxicity are less well understood. A partial listing of factors that influence the cellular response to a cytotoxic agent includes: (1) drug stability and pharmacokinetics under the test conditions; (2) chemical reactions that produce or involve the active moiety, and how these reactions are affected by various culture conditions; (3) intra- and extracellular drug sequestration; (4) the type and site of lesion formed; (5) cellular recovery capability; (6) cell cycle age response; and (7) cell cycle kinetics. All of these factors must be considered in any drug/radiation study. Because chloroethylnitrosoureas (CENUs) are among the most effective drugs against brain tumors, we will briefly review the results of studies on these drugs.

Although substantial progress has been made, except for the effect of some radiosensitizers on cells, no mechanism has been proven for any combination of drug and radiation. This has delayed development of an effective therapy for cancer that is based on combined protocols with drugs and radiation.

Rationale for Combined Radiation Therapy and Chemotherapy

Steel and Peckham [20] have defined four ways in which drugs and radiation can affect tumor cell kill. Spatial cooperation describes the situation where 2 modalities are used to treat disease at different sites. Typically, radiation is used to treat a localized primary tumor and chemotherapy is used to treat metastases. Toxicity independence occurs when both agents have antitumor activity but different dose-limiting normal tissues; for brain tumors, the dose-limiting normal tissue for 1,3-bis(2-chloroethyl)-1-nitrosourea (BCNU) toxicity is unirradiated bone marrow, while the dose-limiting normal tissue for X-rays is the brain. The third category, protection of normal tissues, describes the situation in which the drug used in the combination protocol allows a greater dose of radiation to be delivered to tumor. The fourth category, enhancement of tumor response, will be the focus of this chapter.

Used in combination, drugs and radiation may increase tumor cell kill by one or more general mechanisms described by *Elkind* [4]. Briefly, the first agent may alter the cell cycle distribution of the population by selectively killing cells in a particular phase; as a result, a partially synchronized population would be exposed to the second agent. Total cell kill depends on both the schedule of administration and the efficacy of the second agent. It is also possible that the drug may directly modify repair of the radiation-induced lesions. Radiation may affect membrane permeability such that the amount of drug entering the cell is changed, may directly alter the reaction of the drug with the target molecule, or may interfere with the repair of drug-induced damage. Finally the combined effect of 'sublethal' lesions caused by either of the 2 agents alone may be lethal to cells.

9L Brain Tumor Model

Most advances in cancer therapy have been the result of clinical trial and error. However, experimental studies of the mouse L1210 leukemia model [19] led to the development of a successful treatment for childhood leukemia. A number of experimental models of brain tumors have also been developed, of which the rat 9L model is the most intensively studied [26].

The 9L is a primary brain tumor originally induced in a Fischer rat by N-methylnitrosourea. 9L cells can be grown as monolayer or spheroid cultures and, implanted into rats, can be grown as intracerebral and subcutaneous solid tumors. For animal studies, endpoints are survival and tumor size, and cell survival assays can be used for both in vivo tumors and in vitro cultures. As discussed below, other assays may be used to obtain information that can be used to determine mechanisms. In general, information can be obtained using this model relative to the action of therapeutic agents (the dose-response function), the response of the tumor (recovery mechanisms, resistance and repopulation after treatment), and the influence of the host (drug delivery, dead cell removal and the immune response) that cannot be obtained using a single system.

Single Modalities

Radiation

The action of radiation on mammalian systems is well-defined, and the reader is referred to *Gutin* et al. [6] as well as to other chapters in this book for a more complete description. DNA is probably the most important cellular target of ionizing radiation [7]. High linear energy transfer (LET) radiation such as neutrons can interact directly with biological molecules (direct action); low LET radiation such as X-rays used in therapeutic protocols usually produces an intermediate free radical that chemically disrupts the DNA (indirect action). Both processes produce single and double strand DNA breaks, which are generally believed to be the lesions primarily responsible for radiation-induced lethality. However, radiation-induced DNA strand breaks that may be present in sufficient numbers to be lethal may not necessarily kill cells. Cells have complex, multiple enzyme systems that, over time, can repair these lesions, and lesions that have been correctly repaired have little if any effect on the cell. The balance between the induction of DNA damage and its repair is thought to determine lethality. It is therefore reasonable to assume that it is the unrepaired or misrepaired radiation-induced damage that is responsible for cell death.

Cultured mammalian cells usually exhibit a well-defined cell cycle age response in radiation-induced lethality; cells at the G_1/S border and in M-phase of the cell cycle are the most radiosensitive [9]. These differences in radiosensitivities suggest possible strategies for the combined use of drugs and radiation. If a drug blocks cells in M-phase or at the G_1/S border, increased radiation-induced cell kill would result.

Oxygenation is crucial for the manifestation of radiation-induced damage [8]. Monolayer cells irradiated under oxic conditions are killed more efficiently than cells irradiated under hypoxic conditions. Most spheroids and experimental tumors have a fraction of hypoxic cells; it is generally thought that hypoxic cells may play a role in the radioresistance of many solid tumors.

The radiobiologic response of 9L cells to radiation have been reviewed recently [29]. The radiation survival curve for 9L cells grown in monolayer culture has a much wider shoulder than radiation survival curves for most other cell types, which indicates that 9L cells are relatively radioresistant. This is comparable with the radioresistance of malignant brain tumors, lesions that rarely can be cured by radiation therapy alone. Radiation dose response curves for intracerebral 9L tumors are similar to curves for irradiated 9L monolayer cells. Survival curves show that subcutaneous 9L tumors are significantly more radioresistant than intracerebral 9L tumors or monolayer cells. However, if the subcutaneous tumors are excised, dissociated into single cells, and then irradiated, the survival curve is similar to those of intracerebral tumors and monolayer cells, which suggests that the site of tumor may play a role in the radiosensitivity of tumor cells. Animal survival can be used as an endpoint in these studies, but several factors other than the response of the tumor to radiation may influence the survival of animals with brain tumors [28]; studies that use this endpoint are difficult to design and control because of the extreme complexity of host responses.

It is thought that the recovery from radiation-induced potentially lethal damage (PLD) reflects cellular DNA repair processes. In the 9L system, cells recover from radiation-induced PLD by a factor of 2–3 for cells from exponentially growing monolayers, 10–20 for cells from fed or unfed plateau phase cultures, 10–15 for intracerebral tumors, and 50–100 for subcutaneous tumors [29]. These differences illustrate how the mode of cell growth and/or technical matters can influence recovery of the cells.

Two unusual features of the 9L system are the relatively uniform sensitivity, except for mitotic cells, of monolayer 9L cells to radiation [10, 29] and the lack of an hypoxic fraction in intracerebral and subcutaneous tumors, and in spheroids [3, 29]. The study of drug-radiation combinations in 9L cells may be simplified because synchrony effects with respect to radiation may be disregarded as a possible mechanism. Similarly, because a hypoxic fraction may not exist in the tumor, treatment with a sensitizing drug such as misonidazole would not be considered as a possible mechanism of interaction.

Chloroethylnitrosoureas

The CENUs are a widely used class of antineoplastic agents with established clinical activity against acute lymphocytic leukemia, lymphoma, myeloma, melanoma, gastrointestinal neoplasms, and gliomas. Protocols based on their use, either alone or in combination with radiation, are the most effective treatment known for malignant gliomas. We have studied extensively the effects of these drugs against 9L cells.

Based on data from cell survival curves for 9L cells treated with BCNU, we have developed a drug-dose concept that relates production of the active species to cellular lethality [25]. Drug dose-response curves are considerably more variable than X-ray dose-response curves, primarily because production of the drug species depends critically on treatment conditions such as pH and temperature.

CENUs are considered to be cell cycle nonspecific drugs; however, when 9L cells are sorted by centrifugal elutriation and treated with BCNU, cells in G_1 and in G_2/M are more sensitive than S phase cells (fig. 1). Such anomalies may have possible implications in the development of the optimal sequencing of BCNU and radiation. Numerous in vivo studies using BCNU in the 9L model indicate that the drug kills cells in situ in a dose-dependent manner, and there is no evidence of recovery from PLD [15]. In addition, increasing the BCNU dose increases life span [16].

CENUs decompose rapidly in aqueous media to form a chloroethyl carbonium ion with alkylating activity and a chloroethyl isocyanate with carbamoylating activity. Different CENUs have different alkylating and carbomoylating activities. It is generally thought that while chloroethylation of both DNA and protein occurs, it is alkylation of DNA that is responsible for the cytotoxicity of CENUs. Using alkaline sucrose gradients, *Hilton* et al. [11] found that treatment of L1210 cells with BCNU produced DNA single strand breaks and that the rate and extent of repair was slow and incomplete compared to X-ray-induced single strand breaks. Of more importance are DNA interstrand cross-links that are formed

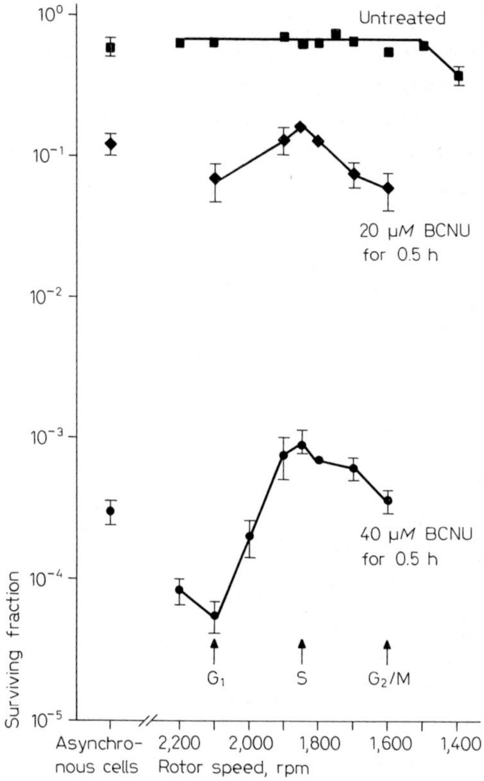

Fig. 1. Survival of 9L cells treated with 0 (■), 20 (◆), or 40 (●) μ*M* BCNU for 0.5 h. Cells were enriched in the various phases of the cycle using centrifugal elutriation. Fractions obtained at various centrifuge rotor speeds are noted that contained mostly G_1, S, or G_2/M cells [redrawn and modified from ref. 33].

subsequent to chloroethylation. DNA cross-link formation is known to increase the cytotoxicity of antitumor agents [5]. Based on data obtained by alkaline elution, *Kohn* [13] postulated that chloroethylation of a DNA base is followed by a slower reaction with a nucleophile on the opposite strand that displaces chloride and forms an ethylene bridge between the 2 strands. In L1210 cells, the maximum cross-link formation measured by alkaline elution occurs 6 h after BCNU treatment. Preliminary experiments performed in our laboratory using alkaline elution indicate a similar time for maximum BCNU cross-link formation in 9L cells.

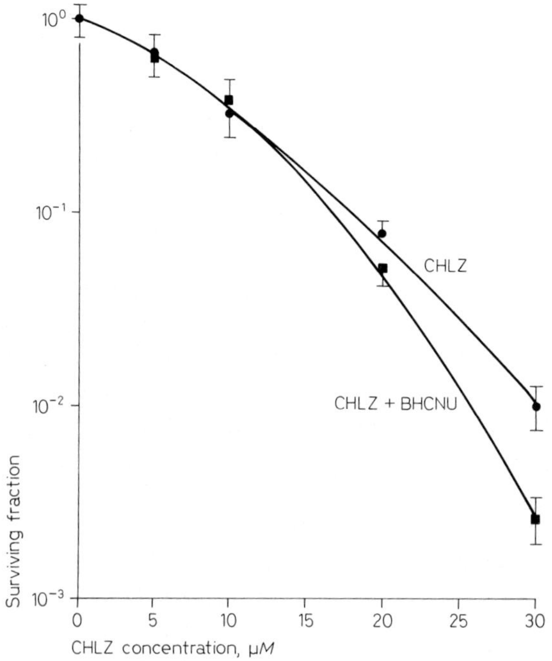

Fig. 2. Survival of 9L cells treated with graded concentrations of CHLZ alone for 1 h (•), or treated with CHLZ and 25 μM BHCNU for 1 h (▪). The curve resulting from the combined drug treatment has been normalized for BHCNU-induced cell kill (SF of BHCNU alone = 0.43).

CENUs may alkylate different sites on DNA, and there are major differences in nitrosourea alkylation between template-active and inactive chromatin and between nucleosomal core particles and linker DNA [1]. In studies of the mechanisms of CENU action, these various reactions must be taken into account.

It is tempting to conclude that CENUs induce cell kill through the formation of DNA cross-links. However, in situations in which both strand breaks and cross-links are produced, alkaline elution data are difficult to interpret because cross-links can mask the production of strand breaks and vice versa. Using viscoelastometry and the S_1 nuclease assay, each of which can detect both single strand breaks and DNA cross-links independently, *Shafer and Chase* [18] detected BCNU-induced single strand breaks in 9L cells, but found no evidence for the production of DNA cross-links. Obviously, more research will be required to clearly define the mechanism of CENU action.

The carbamoylating activity of the isocyanate generated from CENU decomposition, while not considered to be primarily responsible for cytotoxicity, may play a role in the antitumor activity of these compounds. *Kann* et al. [12] performed alkaline elution experiments which suggested that CENUs with carbamoylating activity could inhibit the repair of X-ray-induced single strand breaks. However, these studies used only 1 sequencing protocol and more detailed studies have not been reported. We treated 9L cells in vitro with chlorozotocin (CHLZ), a nitrosourea with alkylating and cross-linking activity, and simultaneously with 1,3-bis(trans-4-hydroxycyclohexyl)-1-nitrosourea (BHCNU), a nitrosourea with carbamoylating activity only. Preliminary results suggest that these 2 nitrosoureas may act synergistically to increase cell kill (fig. 2). *Wheeler* et al. [27] treated mice implanted with L1210 cells with a combination of CCNU, a CENU with alkylating, cross-linking and carbamoylating activities, and CHLZ; mice treated with CCNU followed by CHLZ lived longer than mice treated with either drug alone. The results of these experiments indicate that carbamoylation may be involved in the inhibition of the repair of DNA damage, but they do not prove nor disprove this hypothesis.

Extrapolation of Laboratory Data to the Clinic : An Example

In conventional therapy for brain tumor patients, radiation is delivered in low doses (usually 180 rad) in multiple fractions (usually 30) [17], while BCNU usually is administered in a single high dose (240 mg/m^2) or in a few fairly high dose fractions [23, 24]. When the patient is treated with both modalities, the period between their administration is usually sufficiently long that any possible advantages and/or disadvantages that might result from the combination are negligible.

The possibility of an interaction between BCNU and X-rays was first investigated in the mid-1970s. It was shown in vitro that low concentrations of BCNU could modify the shape of the X-ray cell survival curve [30]; while the final slopes of the combined cell survival curves remained similar to that of X-rays alone, the initial slope was altered considerably by some protocols. A relatively high dose enhancement ratio was pro-

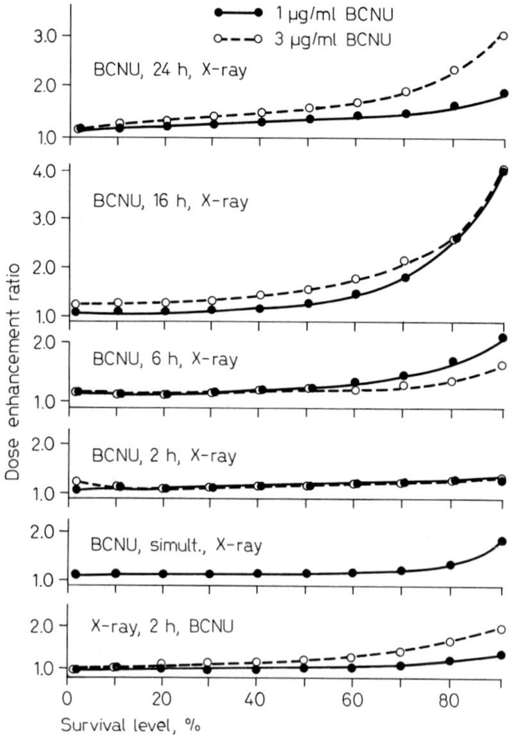

Fig. 3. Dose enhancement ratio obtained by treating monolayer 9L cells with various BCNU-X-ray combinations [redrawn from ref. 30].

duced at high cell survival levels; the highest dose enhancement ratio resulted when BCNU was administered 16 h before irradiation (fig. 3). 95 and 40% of cells treated with 1 and 3 µg/ml of BCNU, respectively, survived drug treatment.

On the basis of these results, in vivo studies were designed to determine whether combinations of BCNU and X-rays might be more efficacious than BCNU alone against the intracerebral 9L tumor. *Barker* et al. [2] showed that single high doses of the 2 agents could be combined to produce a significant increased life span and a modest cure rate in 9L tumor-bearing rats; increase in survival was schedule-dependent. *Wheeler*

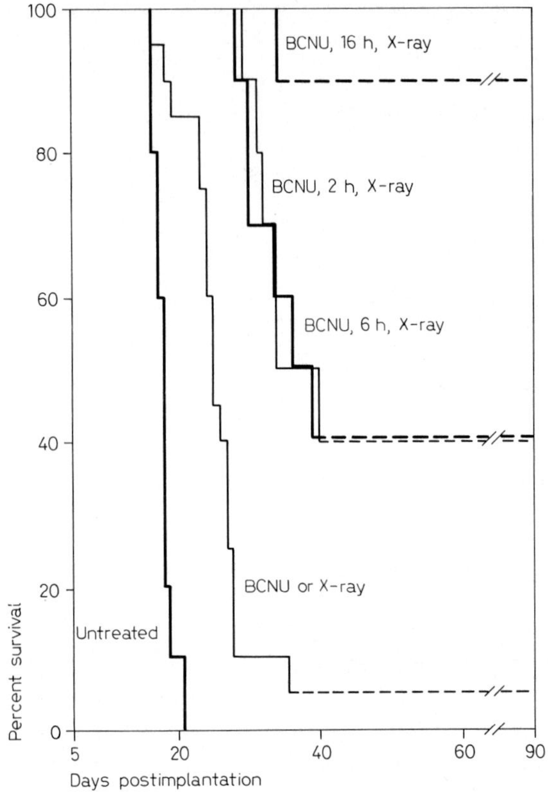

Fig. 4. Animal survival after various BCNU-X-ray combinations. See text for details [redrawn with permission from ref. 31].

et al. [31] administered 10 fractions of X-rays (2,600 rad total) during 2 weeks, with BCNU (2 mg/kg) given either 2, 6, or 16 h before each fraction of radiation. Rats treated with any of these combination schedules lived significantly longer than rats that were irradiated only, and the 16 h schedule was significantly better than either of the other schedules (fig. 4). A lower dose of BCNU (1 mg/kg) in the combined schedules produced no increase in life span.

Thus, the timing of the most effective fractionated dose combination in vivo (BCNU, 16 h before X-rays) was the same as the most effective single dose combination in vitro. While correlations may have been fortuitous given the widely different complexities of the 2 systems, the experimental results suggested that BCNU given to brain tumor patients during radiation therapy might provide an effective treatment. Thus, a combination protocol was designed and tested as a high-risk pilot study for the Eastern Cooperative Oncology Group [17]. 12 patients, 5 with astrocytoma grade III and 7 with astrocytoma grade IV, were entered into this study. The total radiation dose was 7,000 rad administered in five 180-rad fractions/week. The first 6,000 rad was given to the whole brain; the last 1,000 rad was given only to the tumor target volume. The total BCNU dose (240 mg/m^2) was administered as 220 mg/m^2 doses each week 16 h before irradiation. The median survival for the grade III patients was 210 weeks, while the median survival for the grade IV patients was 30 weeks. These numbers were not significantly different from radiation alone controls (median survival for grade III patients, 204 weeks; for grade IV patients, 56 weeks).

One factor contributing to the failure of the protocol was the relatively low BCNU dose used. The pharmacokinetics of BCNU in man were not known at the time of this trial, and the BCNU dose was probably inadequate. We now know that to deliver a dose of 1 µg/ml of BCNU to brain tumor cells in humans would require an intravenous dose of approximately 70–80 mg/m^2 rather than the 20 mg/m^2 used [14]. Because of dose-limiting toxicity, the higher dose could not be administered as often as the lower dose, and any advantage in combination treatment would probably have been lost. However, other, unknown factors besides drug dose may have contributed to the failure of this protocol. Even if the trial were repeated with the higher dose of BCNU, it is doubtful, given the lack of understanding of the mechanism(s) of the effects of drugs and radiation used in combination and therefore our current inability to design a protocol on a sound scientific basis, that the therapeutic effectiveness would improve.

Factors to be Considered in Future Nitrosourea Radiation Studies

To better understand the mechanism of CENU cytotoxicity and the increased cell kill cause by combination treatment with radiation, we have chosen BCNU as a model compound because it has alkylating, cross-linking and carbamoylating activity. We are also studying the effects of nitrosourea analogs that can alkylate and carbamoylate, alkylate and cross-link, or carbamoylate only to define the role of each of these chemical activities in the increased cell kill caused by drugs and radiation.

With data obtained by measurement of gross cellular phenomenona, numerous attempts have been made to explain the combined effects of drug and radiation on the molecular level; clearly, this approach is inadequate. Cell survival, which reflects the combined effects of many factors that lead to cell death, has been used as an experimental endpoint in most studies; the method is most accurate when evaluating treatments that kill at least 50% of cells. Thus, obtaining accurate data at high survival levels is difficult, and assays that measure damage at different subcellular levels should be used. Because CENUs are thought to induce cell death by damaging DNA, we are concentrating on assays that measure DNA damage. It is of course possible to quantitatively determine the amount, site, and repair of this alkylation, but the complexity of the techniques and the long-term nature of the studies make these studies the proper concern of the molecular biologist or biochemist. However, assays such as alkaline elution, viscoelastometry, the sister chromatid exchange (SCE) assay, and the S_1 nuclease assay are more applicable and do provide means to measure gross but relevant effects routinely.

The SCE assay, the use of which was begun recently for the 9L system in our laboratory, is a sensitive, yet relatively simple method for the measurement of damage to DNA and is well-suited to assess the cytotoxic action of CENUs and other cytotoxic drugs. Even though the mechanism of SCE formation is unknown, it is well established that mutagenic and carcinogenic compounds that damage DNA induce SCEs at concentrations far below cytotoxic and clastogenic levels [32].

We have compared results obtained with SCE and colony-forming efficiency (CFE) assays using the 9L model system. The induction of SCE by several nitrosoureas has been directly correlated with their ability to kill cells, and a linear increase in SCE has been observed at nitrosourea doses corresponding to the shoulder region of the survival curves [22]. In addition, 9L cell lines that are resistant to nitrosourea-induced cell kill are also resistant to nitrosourea-induced SCEs [21].

Rigidly controlled experimental conditions eliminate variables and simplify the investigation of complex molecular events. However, the physiological ranges of temperature, pH, O_2 and other nutrients associated with in situ tumors is certainly more variable than those existing in well-controlled cell cultures, but have not been well defined in most experimental tumors and certainly not in the 9L system. Microelectrode studies to determine the pH range of in situ tumors have been begun in our center. In most solid tumors oxygen tension decreases as the distance from the vasculature increases, which has a significant bearing on treatment strategy because of the radioresistance of hypoxic cells. Moreover, because human neoplasms probably consist of cell subpopulations with various degrees of drug sensitivity, the use of homogeneous cell populations is another in vitro culture condition that may confuse the extrapolation of laboratory data to the clinic. Studies of the combined effects of drugs and radiation should be conducted on mixtures of sensitive and resistant cells.

Conclusions

Despite the experimental and conceptual difficulties discussed in this chapter, substantial progress towards an understanding of the ways in which drugs and radiation affect cell kill has been made over the past several years. For instance, preliminary experiments conducted in our laboratory have shown that X-irradiation of 9L cells immediately after treatment for 1 h with 2 μM BCNU produces increases in the number of SCEs/metaphase that are greater than the additive effect of either agent alone. We have tentatively interpreted these data in the following way: Because the half-life of BCNU in medium is less than 30 min, the initial alkylation events have occurred by the time of irradiation; however, radiation may open the helical structure of DNA and make cross-link formation more favorable. This interpretation was suggested by those studies in which 9L cells with low levels of intracellular polyamines were sensitized to BCNU. The absence of polyamines is known to destabilize DNA by reducing the number of supercoiled regions, therefore the increased cytotoxicity of BCNU may result because of an increase in cross-link formation. Studies such as these will, we hope, provide a better understanding of the effect of drugs and radiation on tumor cells.

Acknowledgments

Supported in part by NIH Grants CA-13525, CA-31867, and CA-31868. *P. J. T.* was supported by an Aaron Silvera Fellowship and NIH Training Grant CA-09215. We thank *Neil Buckley* for editing the manuscript.

References

1 Ahlgren, J.D.; Green, D.C.; Tew, K.T.; Schein, P.S.: Repair of DNA alkylation induced in L1210 leukemia and murine bone marrow by three chloroethylnitrosoureas. Cancer Res. *42:* 2605–2608 (1982).

2 Barker, M.; Deen, D.F.; Baker, D.G.: BCNU and X-ray therapy of intracerebral 9L rat brain tumors. Int. J. Radiat. Oncol. Biol. Phys. *5:* 1581–1583 (1979).

3 Deen, D.F.; Hoshino, T.; Williams, M.E.; Muraoka, I.; Knebel, K.; Barker, M.: Development of a 9L multicellular spheroid system and its response to BCNU and radiation. J. natn. Cancer Inst. *64:* 1373–1382 (1980).

4 Elkind, M.M.: Fundamental questions in the combined use of radiation and chemicals in the treatment of cancer. Int. J. Radiat. Oncol. Biol. Phys. *5:* 1711–1720 (1979).

5 Ewig, R.A.G.; Kohn, K.W.: DNA damage and repair in mouse leukemia L1210 cells treated with nitrogen mustard, 1,3-bis(2-chloroethyl)-1-nitrosourea and other nitrosoureas. Cancer Res. *37:* 2114–2122 (1977).

6 Gutin, P.H.; Wara, W.M.; Phillips, T.L.; Wilson, C.B.: Hypoxic cell radiosensitizers in the treatment of malignant brain tumors. Neurosurgery *6:* 567–576 (1980).

7 Hall, E.J.: Radiobiology for the radiologist; 2nd ed., pp.3–12 (Harper & Row, Hagerstown 1978).

8 Hall, E.J.: Radiobiology for the radiologist; 2nd ed., pp.51–62 (Harper & Row, Hagerstown 1978).

9 Hall, E.J.: Radiobiology for the radiologist; 2nd ed., pp.81–94 (Harper & Row, Hagerstown 1978).

10 Henderson, S.D.; Kimler, B.F.; Scanlan, M.F.: Interaction of hyperthermia and radiation on the survival of synchronous 9L cells. Radiat. Res. *92:* 146–159 (1982).

11 Hilton, J.; Bowie, D.L.; Gutin, P.H.; Zito, D.M.; Walker, M.D.: DNA damage and repair in L1210 cells exposed to 1-(2-chloroethyl)-3-cyclohexyl-1-nitrosourea. Cancer Res. *37:* 2262–2266 (1977).

12 Kann, H.E., Jr.; Schott, M.A.; Petkas, A.: Effects of structure and chemical activity on the ability of nitrosoureas to inhibit DNA repair. Cancer Res. *40:* 50–55 (1980).

13 Kohn, K.W.: Interstand cross-linking of DNA by 1,3-bis(2-chloroethyl)-1-nitrosourea and other 1-(2-haloethyl)-1-nitrosoureas. Cancer Res. *37:* 1450–1454 (1977).

14 Levin, V.A.; Hoffman, W.; Weinkam, R.J.: Pharmacokinetics of BCNU in man: a preliminary study of 20 patients. Cancer Treat. Rep. *62:* 1305–1312 (1978).

15 Rosenblum, M.L.; Knebel, K.D.; Vasquez, D.A.; Wilson, C.B.: In vivo clonogenic tumor cell kinetics following 1,3-bis(2-chloroethyl)-1-nitrosourea brain tumor therapy. Cancer Res. *36:* 3718–3725 (1976).

16 Rosenblum, M.L.; Wheeler, K.T.; Wilson, C.B.; Barker, M.; Knebel, K.D.: In vitro evaluation of in vivo brain tumor chemotherapy with 1,3-bis(2-chloroethyl)-1-nitrosourea. Cancer Res. *35:* 1387–1391 (1975).

17 Salazar, O.M.; Van Houtte, P.J.; Bennett, J.M.; Rubin, P.; Wheeler, K.T.: High-dose radiation therapy with low-dose (pulsed) BCNU in malignant gliomas: an eastern cooperative oncology group (ECOG) report. Int. J. Radiat. Oncol. Biol. Phys. *8:* 915–919 (1982).

18 Shafer, R.H.; Chase, E.S.: DNA damage in rat 9L cells treated with nitrogen mustard and 1,3-bis(2-chloroethyl)-1-nitrosourea assayed by viscoelastometry and S_1 nuclease. Cancer Res. *40:* 3186–3193 (1980).

19 Skipper, H. E.; Schabel, F. M.; Wilcox, W. S.: Experimental evaluation of potential anti-cancer agents. XXI. Scheduling of arabinosylcytosine to take advantage of its S-phase specificity against leukemia cells. Cancer Chemother. Rep. *51:* 125–165 (1967).

20 Steel, G. G.; Peckham, M. J.: Exploitable mechanisms in combined radiotherapy-chemotherapy: the concept of additivity. Int. J. Radiat. Oncol. Biol. Phys. *5:* 85–91 (1979).

21 Tofilon, P. J.; Oredsson, S. M.; Deen, D. F.; Marton, L. J.: Polyamine depletion influences drug-induced chromosomal damage. Science *217:* 1044–1046 (1982).

22 Tofilon, P. J.; Williams, M. E.; Deen, D. F.: Nitrosourea-induced sister chromatid exchanges and correlation to cell survival in rat brain tumor cells. Cancer Res. (in press, 1983).

23 Walker, M. D.; Gehan, E. A.: Clinical studies in malignant gliomas and their treatment with the nitrosoureas. Cancer Treat. Rep. *60:* 713–716 (1976).

24 Wasserman, T. H.: The nitrosoureas: an outline of clinical schedules and toxic effects. Cancer Treat. Rep. *60:* 709–711 (1976).

25 Weinkam, R. J.; Deen, D. F.: Quantitative dose-response relations for the cytotoxic activity of chloroethylnitrosoureas in cell culture. Cancer Res. *42:* 1008–1014 (1982).

26 Weizsaecker, M.; Deen, D. F.; Rosenblum, M. L.; Hoshino, T.; Gutin, P. H.; Barker, M.: The 9L rat brain tumor: description and application of an animal model. J. Neurol. *224:* 183–192 (1981).

27 Wheeler, G. P.; Shabel, F. M., Jr.; Trader, M. W.: Synergistic antileukemic activity of combinations of two nitrosoureas. Cancer Treat. Rep. *65:* 591–599 (1981).

28 Wheeler, K. T.: Review of factors influencing the response of brain tumors to therapy. Cancer Treat. Rep. *65:* suppl. 2, pp. 75–81 (1981).

29 Wheeler, K. T.; Barker, M.; Wallen, C. A.; Kimler, B. F.; Henderson, S. D.: Evaluation of 9L as a brain tumor model; in Sridhar, Tumor biology: tissue culture and animal tumor models (Dekker, New York, in press, 1983).

30 Wheeler, K. T.; Deen, D. F.; Wilson, C. B.; Williams, M. E.; Sheppard, S.: BCNU-modification of the in vitro radiation response in 9L brain tumor cells of rats. Int. J. Radiat. Oncol. Biol. Phys. *2:* 79–88 (1977).

31 Wheeler, K. T.; Kaufmann, K.; Feldstein, M.: Response of a rat brain tumor to fractionated therapy with low doses of BCNU and irradiation. Int. J. Radiat. Oncol. Biol. Phys. *5:* 1553–1557 (1979).

32 Wolff, S.: Sister chromatid exchange. A. Rev. Genet. *11:* 183–201 (1977).

33 Bjerkvig, R.; Oredsson, S. M.; Marton, L. J.; Linden, M.; Deen, D. F.: Cell cycle age response of 9L cells to 1,3-bis(2-chloroethyl)-1-nitrosourea and modification by α-difluoromethylornithine. Cancer Res. *43:* 1497–1500 (1983).

D. F. Deen, Ph. D., Brain Tumor Research Center, Departments of Neurological Surgery and Radiation Oncology, School of Medicine, University of California, San Francisco, CA 94143 (USA)

B. Hyperthermia

Prog. exp. Tumor Res., vol. 28, pp. 198–219 (Karger, Basel 1984)

Hyperthermia – Basic Biology[1]

W. C. Dewey, E. V. Holahan

Radiation Oncology Research Laboratory, CED 200, University of California, San Francisco, Calif., USA; Department of Experimental Hematology, Bethesda, Md., USA

Introduction

Hyperthermia has been used as an agent for cancer therapy since the turn of the century. Some encouraging results were obtained, but they were not consistent enough to warrant a sustained effort in developing this application of hyperthermia. During the past few years, however, findings from cell culture studies in animals and from a few preliminary clinical attempts, as well as advances in the design of equipment for delivering hyperthermia, have resulted in some enthusiasm for systematically investigating the potential of using hyperthermia in cancer therapy.

Recent clinical results showing that a significant tumor response is obtained especially when hyperthermia is combined with radiation are encouraging [1, 8, 23]. No definite cures have been reported, but at least 20–30 different groups in the United States and in Europe are involved in clinical investigations of the efficacy of hyperthermia. At the present time, the use of interstitial irradiation and hyperthermia appears most promising [23]. However, future clinical studies need to document clearly that heat combined with radiation obtains better results than are obtained simply by increasing the radiation dose.

The rationale for and problems associated with using hyperthermia in cancer therapy have been reviewed elsewhere [5, 9]. Briefly, enhanced tumor responses should be anticipated because heat kills and sensitizes cells to radiation and chemotherapy. For consistent results, however, the

[1] This research was supported in part by NIH grants (CA-31813 and CA-31808).

temperature and treatment durations must be carefully controlled. This is a particularly critical problem, as adequate thermometry and equipment for heating tumors have not been available – although they are currently being developed. Another advantage of hyperthermia is that heat selectively kills and radiosensitizes those tumor cells that are resistant to ionizing radiation, the cells that are in the S phase of the cell cycle, and those that are hypoxic. The hypoxic cells are most sensitive to heat because they also have low pH levels and are under nutrient deprivation. Another factor that may be important is that heat can prevent recovery from sublethal and potentially lethal radiation damage; this action may or may not be selective for tumor cells. Finally, studies in animals have shown that hyperthermia can improve the therapeutic ratio; i.e., the response of tumor relative to that of normal tissue. However, this therapeutic ratio appears to depend on the sequence in which heat and irradiation are applied, on the number of hyperthermia-radiation fractions, and on physiological changes, such as decreased blood flow, pH, and oxygen concentration associated with hyperthermic treatment. We will consider briefly the rationale, problems, and mechanisms that should be understood in order to go forward systematically with preclinical and clinical studies.

Tumor cells, in particular hypoxic tumor cells at low pH, may develop less resistance to heat than do normal cells existing at physiologically normal pH.

Cell Killing from Heat Alone

Survival curves plotted as the logarithm of survival versus the duration of heating resemble radiation survival curves, in that they have a shoulder followed by a linear portion [5]. The difference, however, is that, after 3–4 h of continuous heating at temperatures of 42.5 °C and lower, thermal tolerance is manifested by the curve bending upward and becoming almost flat (fig. 1). Also, unlike the response to radiation, the response to heat varies by as much as a factor of 10 between different cell lines [5]. Whether or not there is a difference between different cells in different tissues in an individual animal has not been investigated. Furthermore, there is no consistent evidence that tumor cells are more or less sensitive than are normal cells. In most cases, any differences between normal and malignant cells are probably attributable to physiological differences related to pH and nutrient levels.

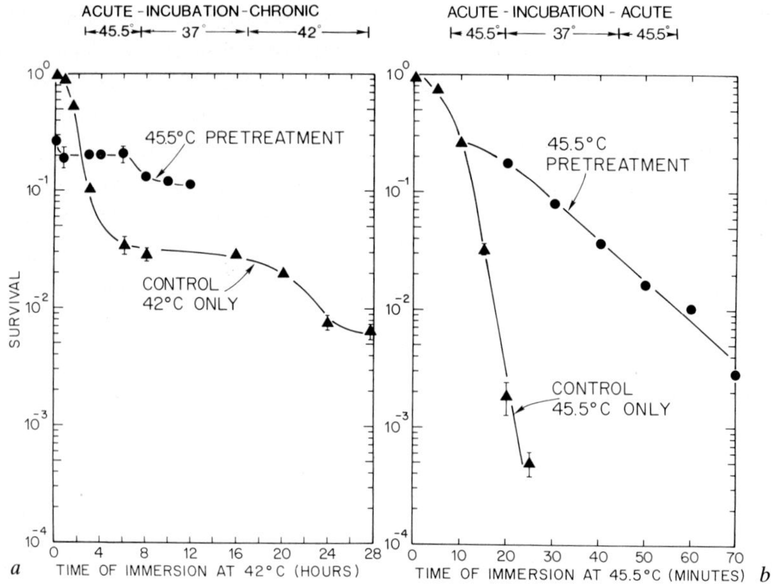

Fig. 1. Thermal tolerance illustrated for prolonged chronic heating and fractionated acute heating. *a* The survival of cells heated at 42°C is shown by the control curve; thermal tolerance is evident after 4 h. The survival of cells preheated for 10 min at 45.5°C, returned to 37°C for 12 h, and heated at 42°C is shown by the closed circles [data from ref. 34]. *b* The survival of cells heated at 45.5°C is shown by the control curve. The survival of cells preheated for 10 min at 45.5°C, returned to 37°C for 12 h, and then heated at 45.5°C is shown by the closed circles.

Cell kill is very sensitive to small variations in temperature [6]. In order to obtain the same response above 42.5°C as that achieved at 42.5°C, for example, a 1°C increase in temperature must be compensated by the duration of heating being reduced by half. Below about 42.5°C, a 0.5°C increase in temperature requires a twofold reduction in the duration of heating. These time-temperature relationships, observed in both cell culture and animal systems, are important in determining the thermal doses delivered at different temperatures. One could attempt to relate all treatments to a standard time at 43°C, for example, if certain fundamental thermodynamic and physiological parameters were known. These conversions to an equivalent thermal dose probably would be most easily established and most consistent for normal tissues, in which the physiological parameters, such as pH, could be predicted more reliably than in tumors.

Cell Killing from Heat Combined with Radiation

Studies have shown that hyperthermia delivered either before, simultaneously with, or after irradiation radiosensitizes mammalian cells, as evidenced by a decreasing shoulder and increasing slope of the radiation survival curve [6]. The greatest sensitization is obtained when hyperthermia and radiation are administered concurrently; the effects are most dramatic when hyperthermia is combined either before or concurrently with irradiation at low dose rates (e.g., 3 rad/min) [3, 12]. This dramatic enhancement of the cells' response to radiation, which is probably related to reduction in recovery from sublethal damage, may be an important rationale for the interstitial application of hyperthermia in combination with low dose rate irradiation administered locally. This combination may be the most fruitful application, especially as the heat can be localized in the tumor quite effectively by using microwave antennae or radiofrequency currents in conjunction with the same implants used for delivering the radiation.

Mechanisms of Hyperthermic Effects

Studies of hyperthermic killing and hyperthermic radiosensitization of mammalian cells [5, 6] indicate that at least two targets are involved, one involving the membranes and the other involving chromatin within the nucleus [4, 33]. Hyperthermia probably has a direct effect on both of these targets. In addition, evidence indicates that damage at the membrane level may lead indirectly to damage at the nuclear level, possibly as a result of changes in the intracellular environment, such as changes in intracellular pH and in the concentrations of certain ions. Although there is much evidence that heat killing and heat radiosensitization involve targets in common [5], there is also evidence that there are separate targets as well. It appears almost certain that the target for radiation killing, which involves direct damage to DNA, is probably different from that involved in either heat killing or heat radiosensitization.

Mechanisms involving physiological changes induced by hyperthermia also are relevant to the application of hyperthermia in cancer therapy. For example, heat frequently decreases the blood flow in tumors while having little effect on, or even causing an increase in, the blood flow in normal tissues [32]. Associated with this decrease of blood flow in tumors is a decrease of extracellular pH and oxygen concentration in tumors. As changes in pH and oxygen concentration can alter the response of cells to radiation and hyperthermia, the mechanisms and conditions associated with these changes should be understood.

Thermal Tolerance

Thermal tolerance is an important factor [14], because recent evidence indicates that it may be used advantageously in applying hyperthermia in cancer therapy. Heating appears to cause the development of resistance to subsequent heat treatments. Thus, the possibility exists that normal tissue at physiologically normal pH may become more tolerant to heat than are tumor cells at low pH.

We wish to emphasize the relationship between thermal tolerance and pH because studies have clearly shown that enhanced hyperthermic killing and radiosensitization are not associated with hypoxia per se, but, instead, with low pH and nutrient deprivation [30]: the oxygen enhancement ratio, which equals the dose to hypoxic cells divided by the dose to oxygenated cells for a specified biological response, is not reduced by hyperthermia, unless the hypoxic cells are also at low pH or are under nutrient deprivation. Therefore, we will consider the possibility of exploiting the phenomenon of thermal tolerance and increased thermal sensitivity at low pH in relation to reducing the oxygen enhancement ratio between oxygenated normal cells and hypoxic tumor cells.

Examples of thermal tolerance in cultured Chinese hamster ovary (CHO) cells are illustrated in figure 1. Figure 1a illustrates how cells become tolerant to continuous heating at 42°C; after about 4 h of heating, no additional decrease in cell survival occurs until tolerance is lost at about 20 h. Furthermore, the curve for 45.5°C pretreatment illustrates that cells heated acutely at 45.5°C for 10 min and then incubated for 12 h at 37°C are tolerant both to subsequent chronic heating at 42°C and to a subsequent acute treatment at 45.5°C (fig. 1b). Other studies [14] have indicated that, when incubation at 37°C continues for 72 h or longer, tolerance to subsequent hyperthermic treatment is lost. Thus, thermal tolerance is not simply a matter of selecting heat-resistant cells, as can be shown directly by studying the heat response of isolated colonies [31]. The mechanisms involved in inducing thermal tolerance are being investigated and, as hypothesized by *Li and Hahn* [21], appear to involve a triggering phenomenon – a development phenomenon involving the synthesis of particular 'heat shock proteins' [22] – and a decay phenomenon. Whereas thermal tolerance, with a reduction in the rate of cell inactivation, can develop at either 37 or 42°C after thermal shock (fig. 1), a short acute treatment at 45.5°C greatly increases the rate of cell inactivation, if the acute treatment is delivered immediately before chronic exposure to the low temperature of 42°C [34]. This type of sensitization and the protocol has been referred to as 'step down heating' [13]. Furthermore, because of the phenomenon of thermal tolerance, 'step up heating' – chronic heating at low temperatures followed by acute heating at a high temperature – can give a 100-fold higher survival level than can step down heating [6].

Heat Killing at Acidic versus Physiologically Normal pH: Heating under Nontolerant Conditions or as Thermal Tolerance is Developing

When CHO cells were heated at 45.5°C, no thermal tolerance developed, as was expected (fig. 2). Furthermore, lowering the pH sensitized the cells to hyperthermia by reducing the shoulder of the survival curve. However, when the cells were heated chronically for several hours at 42°C, tolerance developed under both pH conditions. The important observation is that the curve for the acidic pH diverges more from the curve at the physiologically normal pH when cells are heated chronically at 42–42.5°C than when they are heated acutely at 45.5°C. This difference is probably the result of tolerance developing more slowly under acidic conditions than under physiologically normal pH conditions. The great enhancement in heat killing at low pH levels that was observed during chronic heating at 42–42.5°C means that, compared to cells at the physiologically normal pH values that are heated under the same time and temperature conditions (i.e., with the same thermal dose) required to reduce survival to some value 10%, cells heated under acidic conditions will have a much lower survival level from chronic heating at 42–42.5°C than from acute heating at 45.5°C.

If these observations in vitro are applicable to tumor cells at low pH, then compared with normal cells at physiologically normal pH, one would expect a greater thermal gain (i.e., tumor kill relative to normal tissue damage) for heating under chronic conditions at 42–42.5°C than for acute heating at 45.5°C.

Similar experiments carried out with fractionated acute heat treat-

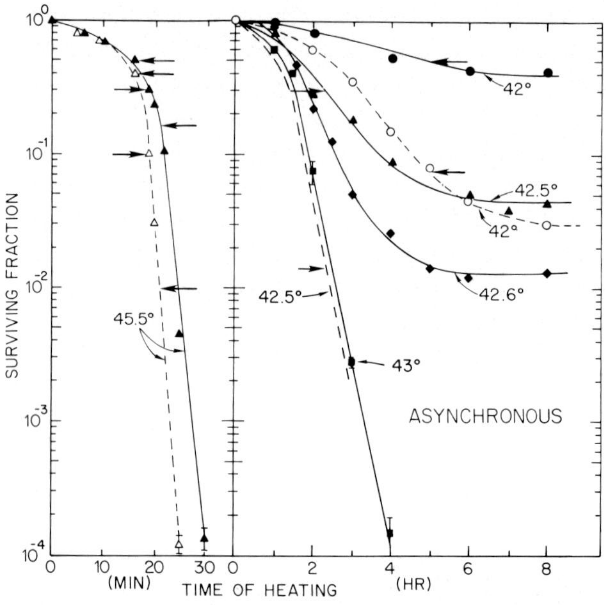

Fig. 2. Fraction of asynchronous CHO cells surviving hyperthermic treatment as a function of temperature, duration of heat treatment, and extracellular pH. Cells were exposed at pH 7.4 (closed symbols) or 6.75 (open symbols). The effect of acidic pH on hyperthermic sensitivity was compared at the 0.50 (←), 0.30 (→), and 0.16 (←) surviving fraction levels for pH 7.4.

ments separated by 6–12 h indicate that less tolerance develops in cells at low pH [11] and under nutrient deprivation [24] with little effect of hypoxia itself [25]. Additional data are needed, however, concerning the kinetics, magnitude, and decay of tolerance under the two pH conditions.

Heat Radiosensitization for Cells Treated at pH 6.75 or 7.4

Data indicate that as tolerance for heat killing developed, tolerance also developed for heat radiosensitization when CHO cells were heated chronically for several hours at 42°C under acidic conditions (fig. 3). Note that although tolerance was manifested by a decreased slope, no tolerance was evident for the shoulder of the curve. A similar observation was reported [15, 16] for fractionated doses at normal pH. Furthermore, as the cells were heated acutely at 45.5°C, tolerance did not develop and the

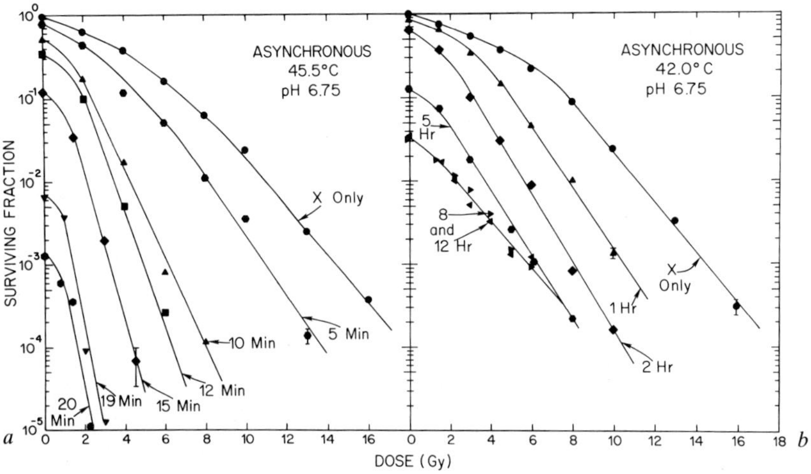

Fig. 3. Heat-induced hyperthermic radiosensitization in asynchronous CHO cells. Cells were heated at 45.5°C for 5–20 min *(a)* or at 42°C for 1–12 h *(b)* and then incubated at 37°C for 10 min before X-irradiation. The cells were heated and irradiated at acidic pH. When cells were heated and irradiated at pH 7.4, the curves were similar except for less thermal damage; i.e., at 45.5°C 17 min corresponded to 12 min at pH 6.75, and at 42°C 5 h corresponded to 2 h at pH 6.75. However, for radiation only, killing was greater at pH 7.4 than at pH 6.75, i.e., 10^{-4} versus 10^{-3} survival.

amount of heat radiosensitization increased as the amount of heat killing increased. An additional observation was that low pH enhanced survival in a situation of irradiation without any heat treatment (table I). This finding has been reported previously [17]; it has been attributed to the low pH after irradiation either enhancing recovery or decreasing fixation of radiation damage. Many similar experiments carried out with G_1 and asynchronous CHO cells indicate that the results depicted in figure 3 represent qualitatively the results for treatments either at pH 6.75 or 7.4; the results differ quantitatively, however, in that the amount of heat killing and heat radiosensitization for a specified time and temperature are greater at the low pH, as compared to the basic pH. However, the survival curves and absolute survival levels are virtually the same for the two pH conditions if the temperature under basic conditions is about 0.5°C higher than the temperature under the acidic conditions.

An important question of clinical interest is how the pH conditions existing in the tumor at the time of heating and irradiation influence the

Table I. Effect of temperature and pH on the TER[1] of asynchronous cells

Heat treatment	pH 7.40 surviving fraction heat only	X-ray[2] dose, Gy	TER_{HR}	TER_R	TER_H	pH 6.75 surviving fraction heat only	X-ray[2] dose, Gy	TER_{HR}	TER_R	TER_H
42°C										
2 h	0.80	8.6	1.38	1.35	1.02	0.60	9.4	1.51	1.44	1.05
3 h	0.70	8.4	1.42	1.38	1.03	0.35	8.2	1.74	1.56	1.12
5 h	0.50	7.7	1.55	1.45	1.07	0.08	5.5	2.58	1.93	1.34
8 h	0.40	8.7	1.36	1.25	1.09	0.03	5.2	2.73	1.76	1.55
42.5°C										
1 h	0.74	8.3	1.43	1.40	1.02	0.50	6.6	2.15	2.01	1.07
1.5 h	0.50	6.5	1.84	1.72	1.07	0.25	4.3	3.38	2.82	1.17
3.5 h	0.25	5.3	2.24	1.95	1.15	0.01	1.4	10.50	5.59	1.88
45.5°C										
10 min	0.70	8.2	1.44	1.39	1.04	0.70	8.8	1.62	1.57	1.03
16 min	0.50	6.7	1.77	1.66	1.07	0.40	6.9	2.08	1.89	1.10
19 min	0.30	5.1	2.32	2.04	1.14	0.10	3.6	3.95	3.03	1.30
20 min	0.23	4.8	2.48	2.11	1.18	0.03	2.0	6.97	4.48	1.56
21 min	0.16	3.9	3.05	2.48	1.23	0.01	1.1	12.91	6.87	1.88

[1] As illustrated in figure 4, TER_{HR} is for both heat killing and heat radiosensitization, TER_R is the contribution due to heat radiosensitization, and TER_H is the contribution due to heat killing. The TER values and X-ray doses (tabulated for $\Delta+X$) were determined at an isosurvival level of 10^{-3}.
[2] The X-ray doses for radiation only were 11.9 and 14.2 Gy for pH 7.4 and 6.75, respectively.

amount of cell killing for a specified time-temperature combination. For such an analysis, curves similar to those plotted in figure 3 were analyzed according to the schematic model of survival curves illustrated in figure 4. Curves have been constructed to illustrate increased cell killing resulting from both heat killing and heat radiosensitization (HR), from heat killing by itself (H), and from heat radiosensitization by itself (R). Quantitatively, thermal enhancement ratios (TERs) were determined by the ratios of the doses to give an isoeffect level of 10^{-4} or 10^{-3} survival. The thermal gain factor (TGF) was then determined by the ratio of the TER at pH 6.75 divided by the TER at pH 7.4.

The TERs and TGFs for treating at pH 6.75 relative to treating at pH

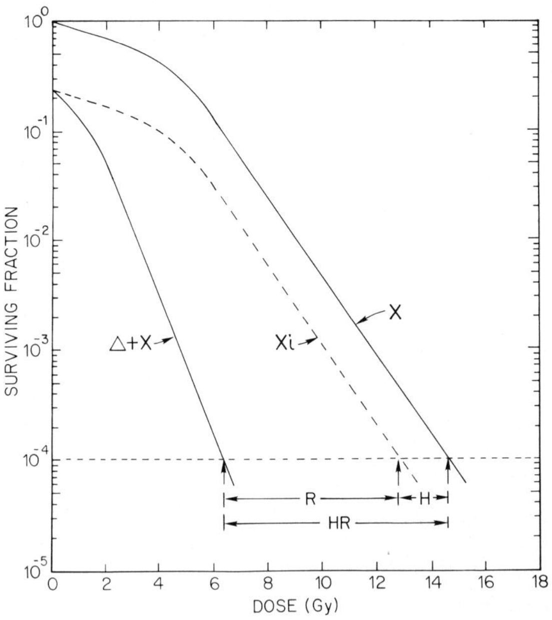

Fig. 4. Schematic representation illustrates how TER is the result of both heat killing and heat radiosensitization. The curve X is for radiation only, and the curve Δ+X is for heat combined with radiation. The curve X_i is drawn parallel to the curve X to illustrate the survival for heat combined with radiation, if heat killed cells without any radiosensitization. Therefore, at a specified survival value (10^{-4} for example), TER_{HR} for both heat killing and heat radiosensitization is $D_X/D_{\Delta+X}$; TER_H for heat killing only is D_X/D_{Xi}, and TER_R for heat radiosensitization only is $D_{Xi}/D_{\Delta+X}$. Thus, $TER_{HR} = TER_H \times TER_R$. Then, to obtain a TGF, for which the survival curves for pH 6.75 are compared with those for pH 7.4, the TGF = TER for pH 6.75/TER for pH 7.4.

7.4 are illustrated in table I and figure 5. The TGF is plotted relative to the survival fraction observed for heat killing only at pH 7.4. For example, a surviving fraction of 0.5 would specify a particular time-temperature combination (e.g., 1.5 h at 42.5 °C) required to reduce the survival to 0.5 for heat treatment only at pH 7.4. Cell survival then was determined for both heating alone and heating combined with irradiation, when the cells were treated under the same time-temperature combination at pH 6.75 or 7.4. In table I, the duration of heating versus three temperatures are indicated as a function of survival at pH 7.4. The data for TGFs in figure 5 indicate, as expected from the survival data in figure 2, that the TGF for a specified amount of heat damage at pH 7.4 is greater for heating at 42–42.5 °C than

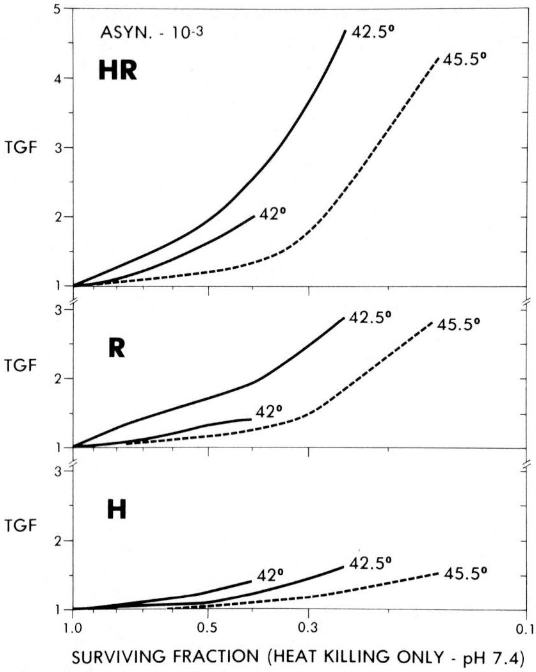

Fig. 5. The variation in TGF of asynchronous CHO cells with increased hyperthermic killing. The TGF (defined in the legend of figure 4) for cells heated at 42, 42.5 or 45.5°C was calculated from TER values at a survival of 10^{-3} for pH 6.75 and 7.4 (table I). The variation in the TGF due to hyperthermic killing plus hyperthermic radiosensitization (HR), hyperthermic radiosensitization (R), and hyperthermic killing (H), was plotted against the fraction of cells surviving the heat treatment only at pH 7.4. Table I shows the durations of heating required to give a survival of 0.5, for example, i.e., 5 h at 42°C, 1.5 h at 42.5°C, and 16 min at 45.5°C.

for heating at 45.5°C. Although the TGFs were determined with asynchronous cells for an isoeffect of 10^{-3}, similar results were obtained at 10^{-4} and with synchronous G_1 cells. Furthermore, a comparison of these data with the data in figure 2 for heat killing only shows that the therapeutic gain does not begin to increase until significant cell killing occurs from the hyperthermic treatment alone. Also, the increase in the TGF is attributable much more to an increase in radiation sensitization at low pH than to an increase in heat killing at low pH. Nevertheless, there is some increase in the TGF that can be attributed to heat killing only. This gain may be of importance in a protocol in which hyperthermia is delivered several hours or days before or after radiation treatment.

From the TERs for R shown in table I, note that the thermal gain from heat radiosensitization is associated much more with differential heat killing at the acidic pH than with differential thermal tolerance for heat radiosensitization, which is seen only for 8 h at 42 °C. Also, note that the data in table I and figure 5 are based on radiation doses low enough to be relevant to fractionated doses used in the clinic.

Relation of Observations in vitro to Observations in vivo

Experiments have indicated that occluding the blood supply to normal tissue, which should lower the pH, enhances the response to hyperthermia [2, 36] and that this enhancement is more pronounced at temperatures of about 42–43 °C than those of 45.5 °C or above [2]. These observations in vivo are exactly what would be expected from observations in vitro (fig. 2, 5). Furthermore, as the pH of blood can be lowered with a concomitant selective lowering of pH in the tumor relative to normal tissue by infusing animals with glucose [20, 38], a selective increase in the hyperthermic sensitivity of tumors should be observed as the pH is lowered in vivo [20, 38]; this has proved to be the case. Moreover, the enhancement associated with low pH was greater for treatment at 42 °C than for treatment at 45.5 °C [38]. These observations in vivo correlate very well with observations in vitro (fig. 2, 5) and support the contention that low pH observed in at least some tumor cells may be relevant to applications of hyperthermia in cancer therapy.

Studies in vivo show that thermal tolerance occurs in cases of fractionated heat treatments that are not combined with irradiation [14, 18, 39, 40] and that it may occur in cases of heat radiosensitization as well [9]; i.e., the TER for heat radiosensitization of normal tissue decreased relative to an increase in the number of radiation fractions when 43 °C for 30 min immediately preceded each radiation dose. This change, however, may have been mainly a consequence of less heat radiosensitization associated with less heat killing (damage) as thermal tolerance developed, instead of with thermal tolerance for heat radiosensitization itself.

Studies of the effects of hyperthermia combined with irradiation on the cure of mouse tumors also should be considered in relation to our in vitro studies. Pertinent results of in vivo studies by three investigators [26,

Table II. Effect of 1 h hyperthermia after irradiation on cure of C3H mammary carcinoma[a]

Heat treatment	TCD-50 O_2, Gy	TCD-50 air, Gy
None	–	63.5[b]
None	–	56.3[c]
None	55.1[d]	59.5[d]
Misonidazole	–	25.7[c]
41.5°C	–	53.1[c]
44°C for 15 min = 42–42.5°C	27.0[d, e]	42.7[d, e]
42.5°C	–	36.9[c]
Misonidazole + 42.5°C	–	24.3[c]
42.5°C	–	36.8[b]
Misonidazole + 42.5°C	–	19.4[b]
43.5°C	–	26.7[c]
Misonidazole + 43.5°C	–	18.4[c]

[a] TCD-50 values are listed for radiation only, two atmospheres of hyperbaric oxygen at the time of irradiation (O_2), hyperbaric oxygen at the time of irradiation plus hyperthermia (with air) after irradiation, radiation plus hyperthermia, misonidazole (1 mg/g) 30 min before irradiation, and misonidazole with irradiation plus hyperthermia. The relative number of viable cells in each compartment of the tumor is about 60% oxygenated, 30% acutely hypoxic, and 10% chronically hypoxic [7, 37]. See text for definitions and references.
[b] Data from *Stone* [36], heat about 5 min after irradiation.
[c] Data from *Overgaard* [26, 28], heat 4 h after irradiation.
[d] Data from *Thrall* et al. [37], heat about 5 min after irradiation.
[e] These data were actually obtained at 44°C for 15 min which should give the same thermal damage as 42–42.5°C for 60 min (see text).

28, 36, 37] are summarized in table II. From the doses required to cure 50% of the tumors (TCD-50), the effects on oxygenated, acutely hypoxic, and chronically hypoxic cells in the tumor can be estimated [7, 28]. Acutely hypoxic cells have been defined as those cells that can be oxygenated with hyperbaric oxygen; chronically hypoxic cells are defined as those that cannot be oxygenated. Oxygenation is determined by a cell's radiosensitization with hyperbaric oxygen [7, 37]. As illustrated qualitatively in table II, a single dose of radiation kills most of the oxygenated cells, but leaves most of the acutely and chronically hypoxic cells undamaged. The acutely hypoxic cells, however, can be killed if the tumors are irradiated while the animal breathes hyperbaric oxygen. To kill the chronically hypoxic cells by radiation therapy, irradiation must be combined with some other treatment, such as misonidazole or hyperthermia for 1 h at 42.5–43.5 °C. We are interested that, in order to kill the acutely hypoxic cells by combining heat with radiation, the heating damage must be increased to a level greater than that required to kill the chronically hypoxic cells; i.e., from 42.5 °C for 1 h to 43.5 °C for 1 h. This might be expected from the in vitro data that show that significant heat killing must occur in order to obtain a selective effect on cells with low pH (fig. 2). However, there is a good possibility that many of the acutely hypoxic tumor cells may not be very acidic, relative to the oxygenated tumor and normal cells. This possibility needs to be investigated. Also of interest is the observation that the TCD-50 values for misonidazole combined with irradiation are about the same as those for hyperthermia at 43.5 °C combined with irradiation. Of course, the greatest cell killing is obtained with the combination of irradiation, hyperthermia, and misonidazole because of the selective effect of misonidazole on hypoxic cells [26]. Because neither misonidazole before irradiation nor heat delivered 4 h after irradiation enhanced the effect of radiation on normal skin, a therapeutic gain of 2 for heat and 3 for misonidazole plus heat was observed [26].

Certainly, as indicated by the in vitro results (fig. 2, 5) and the in vivo results (table II), the radioresistant hypoxic cells at low pH and under nutrient deprivation should be selectively sensitized to heat killing and heat radiosensitization. Also, they should develop less resistance or tolerance to heat. Thus, the chronically hypoxic tumor cells should definitely show a thermal gain from hyperthermic treatment, even if they are at the same temperature as the cells in limiting normal tissue. However, the oxygenated and acutely hypoxic cells in the tumor may be at physiologically normal pH and may have a sufficient level of nutrients and, thus, may not be selectively killed or radiosensitized by hyperthermia. Therefore, to insure a thermal gain, either the oxygenated tumor cells should be raised to a higher temperature than the normal cells, or possibly the pH of the oxygenated tumor cells could be lowered by means of intravenous injections of glucose [20, 38], conceivably combined with insulin. Nevertheless, the numerous oxic cells may indeed limit the effectiveness of radiation therapy combined with hyperthermia.

In fact, in vivo data comparing the therapeutic gains for single heat treatments with those for fractionated heat treatments appear to support our suggestion that oxygenated cells in tumors, especially with reoxygenation between fractions, may be just as important to the success of therapy as, or more important than, hypoxic cells at low pH. In other words, if oxygenated cells are most important, no increase in therapeutic gain would be expected as the thermal treatment is fractionated. For example, *Stewart and Denekamp* [35] observed a therapeutic gain of 1.2–1.5 for tumor growth delay (TER = 1.2–1.5) compared to skin response (TER = 1.0) when one dose of radiation was followed 3 h later by heating (1 h at about 42.5°C); however, the therapeutic gain decreased to 0.9–1.0 when the same radiation-heat protocol was used for treatments delivered in either two or five fractions, with each fraction separated by 24 h. Also, *Urano* et al. [39] observed only a small increase (about 14%) in the growth delay of mouse tumor compared to the response of normal tissue (toe loss) when the results of two or three heat treatments (without irradiation) separated by 48 h were compared with those of a single heat treatment. Thus, the thermal tolerance they observed did not appear to have a large selective effect on tumor cells, unlike what might be predicted from in vitro and in vivo results mentioned previously. But these data and similar data of *Overgaard* et al. [29] were based on dose-response curves for heating alone (43.5°C for 1–7 h), from which it would seem that hypoxic tumor cells at low pH should be selectively killed by heat, leaving the oxygenated cells (and conceivably some acutely hypoxic cells) at the physiologically normal pH to determine the growth delay and cure rates of tumors. In addition, other complications may exist in these studies. For example, with water bath heating [35], deep portions of the tumor may be as much as 1.5°C lower than the temperature of the skin. Furthermore, in the *Stewart and Denekamp* [35] studies, there was no thermal enhancement of the radiation effect on skin, so that selective thermal tolerance in skin relative to tumor could play a role. However, even when the TER in skin was increased to 1.7–2.5 [27, 35] by irradiating either during heating (42.5°C for 1 h) or immediately thereafter, no evidence of more thermal tolerance in skin than in tumor was observed with two or five fractions, with 24 h between each fraction: there probably was not enough heat killing after the first fraction for any thermal tolerance, which could be manifested by less heat killing in skin than in tumor, to give a significant reduction in the TER of skin relative to the TER of tumor. In fact, if thermal tolerance developed between fractions in both tumor and normal tissue, but to a lesser extent in tumor than in normal tissue, a thermal gain owing to differences in pH, as shown by our in vitro results, would not be expected unless significant heat killing occurred (fig. 2, 5). Therefore, before conclusions can be reached concerning the ineffectiveness of multifractions to increase the thermal gain because of reduced tolerance in hypoxic tumor cells, heat treatments should be increased with each fraction as thermal tolerance develops in order to obtain significant heat killing and a significant TER with each radiation-heat fraction.

As the question of selective thermal tolerance is considered, hyperthermia combined with irradiation should be emphasized. For example, if radiation were combined with heat in order to kill the oxygenated cells, the hypoxic cells at low pH might still respond selectively to heat, even in fractionated protocols. In that case, with significant heat killing, the development of thermal tolerance during fractionation might have a selective effect on sparing normal tissue more than tumor. An interesting observation [27] related to this discussion was that the TER for tumor from five fractions separated by 24 h was as great (1.35) when heat (42.5°C for 1 h) was delivered 3 h after the last fraction as when it was delivered 3 h after each fraction (TER = 1.37, which gave a TGF of 1.4). This strongly suggests that either thermal tolerance reduced the effectiveness of the heat doses, or because of partial reoxygenation between radiation fractions, the selective effect of hyperthermia on chronically hypoxic cells is only important for those hypoxic cells remaining after the last fraction is given. Further experiments are needed to address these questions, including the possibility that thermal tolerance, which may persist for as long as 2 weeks [40], may decay more rapidly in tumor than in normal tissue, as was suggested [39] by an increase in thermal gain when the interval between each of two or three heat treatments was increased to 5 days. This may be particularly relevant for dividing brain tumor cells as compared to nondividing normal brain cells because there is some evidence that loss of tolerance may be associated with cell division.

Importance of the Sequence between Hyperthermia and Radiation Relative to Physiological Changes in the Tumor Induced by Hyperthermia

The studies of *Song* et al. [32] and others have shown that hyperthermia treatments can reduce the blood flow, oxygen concentration, and pH in tumors relative to those in normal tissues. In fact, in normal tissues, the blood flow frequently increases after hyperthermic treatment. *Song* et al.

[32] showed in rats that a treatment of 43.0°C for 1 h can significantly lower the pH in tumor and that a second treatment following the first can lower the pH even more. In these same studies, *Song* et al. [32] showed that the blood flow also decreased. Thus, after a hyperthermic treatment, the decrease in pH and associated decreases in blood flow and oxygen concentrations may have significant effects on the response of the tumor cells to hyperthermia and/or radiation. Data of this type are needed from studies of both brain tumors and normal brain tissue.

Studies in vitro have shown that the greatest amount of cell killing is obtained when radiation is delivered concurrently with the hyperthermic treatment [5, 6]. At the physiologically normal pH of 7.4, irradiation either before or after the hyperthermic treatment is less effective than is simultaneous treatment with both modalities. The same pattern is observed at pH 6.75, except that survival is lower at this pH level than at pH 7.4; and when heat precedes irradiation and the pH is kept low, the thermal damage remains for several hours to interact with the radiation dosage. Results in vivo in mice [5] and in humans [27] agree with those in vitro for a simultaneous treatment and for irradiation preceding the hyperthermic dose. When hyperthermia precedes the radiation treatment [5, 9, 35], the pattern for normal tissue is similar to that observed in vitro at pH 7.4. But the effect on tumor is variable, and the thermal damage generally does not persist as it did for cells treated in vitro at pH 6.75. This lack of a definite thermal gain when heat precedes irradiation is thought to be associated with the physiological changes discussed earlier. Although these studies can be criticized because portions of the tumor may have been at a lower temperature than the normal tissue (skin in most cases) [35], only the absolute TER and TGF and not the general relationship between TER and sequence should be affected. Thus, based on in vitro results, heating followed by irradiation would be expected to give the greatest thermal gain or therapeutic ratio in vivo, but this was not observed, probably because heat lowers blood flow, pH, and the oxygen concentration in the tumor, thus increasing the radioresistant hypoxic cell fraction. Such an increase in the hypoxic fraction is probably most important in terms of tumor cures from large single doses of radiation and may not be so important for fractionated doses or for heating preceding low dose rate interstitial irradiation. Nevertheless, as results in vitro are compared with those in vivo, physiological changes must be considered, especially as hyperthermia and radiation are delivered in multifraction protocols.

Considerations Important for Applying Hyperthermia to Cancer Therapy

The possible advantages of hyperthermia in cancer therapy include the following. (1) The radioresistant S phase cells in a growing tumor should be selectively sensitized to heat killing and heat radiosensitization which may provide a therapeutic gain compared to that for noncycling normal cells. (2) The radioresistant hypoxic tumor cells that are nutritionally deficient and at low pH should be selectively sensitized to heat killing and heat radiosensitization. (3) There may be different repair kinetics in tumor and normal tissue for radiation delivered before or after hyperthermia, especially as influenced by conditions of pH. (4) Because of poor blood circulation in tumor relative to that in normal tissue, there is a possibility of obtaining a higher temperature in the tumor than in normal tissue. This method is actually being exploited clinically in treating large tumors and may be used to advantage in selectively enhancing the effect of chemotherapeutic agents on tumors. Because hyperthermia complements radiation in terms of selectively killing radioresistant S phase and hypoxic cells, hyperthermia should be considered as a modality to be combined with irradiation.

Before hyperthermia can be used most effectively in the clinic, however, several problems remain to be resolved. These include: mechanisms of heat killing and heat radiosensitization, the best temperature and duration to use, the best sequence for applying heating and irradiation, the best interval between heat and X-ray fractions (to take advantage of possible differences in thermal tolerance), how to heat the tumor and adjacent normal tissue with precise temperature control (probably the most limiting factor presently), the possible spreading of metastases that could be associated with hyperthermic effects on immunological processes, and understanding physiological changes discussed previously.

As these factors are investigated, an important consideration is the possible difference between heat and radiation in terms of the time at which the damage to normal tissue occurs after treatment. As observed by some workers [9, 19; *Stone,* personal commun.], damage to normal tissue from heat alone appears to occur early, in contrast to critical late effects from radiation. Certainly, if heat radiosensitization occurs from heat damage interacting with radiation and thus enhancing the radiation response, late radiation damage in normal tissue should be increased [10]. However, an important question is: if heat is separated from radiation such that it is not interacting with radiation in terms of heat radiosensitization, will late effects that limit the amount of heat and radiation that can be delivered be enhanced? Thus, studies are needed in which damage is assayed early and late after hyperthermia and radiation treatments are delivered in either the interactive or noninteractive mode. In the interactive mode, a therapeutic gain will be obtained only if the radiation dose-response curve for tumor cure is shifted more than the radiation dose-response curve for normal tissue complications. However, in the noninteractive mode, a therapeutic gain may be obtained simply by shifting the radiation dose-response curve for tumor cures, with little effect on the dose-response curve for normal tissue complications.

As various fractionation protocols are investigated for hyperthermia combined with irradiation, either in the interactive or noninteractive mode, special attention should be focused on the amount of hyperthermia that can be delivered without exceeding the tolerance of limiting normal tissue. Therefore, data must be obtained for normal tissue damage as a function of time-temperature combinations when heat is delivered with radiation, in order that the equipment delivering hyperthemia can be controlled to deliver the maximum amount of heat based on approaching, but not exceeding, the tolerance to limiting normal tissue. Obviously, time-temperature combinations will have to be evaluated in order to establish standard hyperthermia doses to normal tissue. For this reason, as results in animals and humans are obtained, the temperatures and durations of heat treatments in the normal tissue that is being exposed to radiation and hyperthermia must be recorded accurately. Also, as the heat is focused as much as possible in the tumor in order to deliver a greater thermal dose to the tumor than to the normal tissue, the time-temperature profile for the tumor must be accurately recorded. Optimally, the minimum temperature in the tumor will exceed the maximum temperature in the normal tissue.

A critical factor in applying hyperthermia to cancer therapy is to control the heat distribution such that any cold spots in the tumor are no lower in temperature than any hot spots in normal tissue, because cold spots in the tumor will undoubtedly contribute to failures from hyperthermia treatments. Note that the TGFs reported in this chapter for the pH-sensitizing effect, exclusive of any differences in thermal tolerance with fractionated doses at different pH values, would be eliminated if the temperature of the cells at low pH were only about 0.5°C lower than the temperature of the cells at normal pH. Thus, as the physical parameters for time and temperature determinations of thermal doses in normal tissue and tumor are accurately recorded, normal tissue complications, tumor cures or local control, and patient survival must be documented. A most important consideration is to identify the factors associated with both obtaining and failing to obtain a therapeutic gain.

References

1 Arcangeli, G.; Barni, E.; Cividalli, A.; Mauro, F.; Morelli, D.; Nervi, C.; Spano, M.; Tabocchini, A.: Effectiveness of microwave hyperthermia combined with ionizing radiation: clinical results on neck node metastases. Int. J. Radiat. Oncol. Biol. Phys. *6:* 143–148 (1980).

2 Baker, G.M.; Waas, A.N.C.; Wright, E.A.: The influence of ischaemia on hyperthermic damage to the mouse tail. Int. J. Radiat. Biol. *37:* 109–114 (1980).

3 Ben-Hur, E.; Elkind, M.M.; Bronk, B.V.: Thermally enhanced radio-response of cultured Chinese hamster cells: inhibition of repair of sublethal damage and enhancement of lethal damage. Radiat. Res. *58:* 38–51 (1974).

4 Clark, E.P.; Dewey, W.C.; Lett, J.T.: Recovery of CHO cells from hyperthermic potentiation to X-rays: repair of DNA and chromatin. Radiat. Res. *85:* 302–313 (1981).

5 Dewey, W.C.; Freeman, M.L.; Raaphorst, G.P.; Clark, E.P.; Wong, R.S.L.; Highfield, D.P.; Spiro, I.J.; Tomasovic, S.P.; Denman, D.L.; Coss, R.A.: Cell biology of hyperthermia and radiation; in Meyn, Withers, Radiation biology in cancer research, pp. 589–621 (Raven Press, New York 1980).

6 Dewey, W.C.; Hopwood, L.E.; Sapareto, S.A.; Gerweck, L.E.: Cellular responses to combinations of hyperthermia and radiation. Radiology *123:* 463–474 (1977).

7 Dewey, W.C.; Thrall, D.E.; Gillette, E.L.: Hyperthermia and radiation. A selective thermal effect on chronically hypoxic tumor cells in vivo. Int. J. Radiat. Oncol. Biol. Phys. *2:* 99–103 (1977).

8 Dritschilo, A.; Piro, A.J.: Therapeutic implications of heat as related to radiation therapy. Semin. Oncol. *8:* 83–91 (1981).

9 Field, S.B.; Bleehen, N.M.: Hyperthermia in the treatment of cancer. Cancer Treat. Rev. *6:* 63–94 (1979).

10 Goffinet, D.R.; Choi, K.Y.; Brown, J.M.: The combined effects of hyperthermia and ionizing radiation on the adult mouse spinal cord. Radiat. Res. *72:* 238–245 (1977).

11 Goldin, E.M.; Leeper, D.B.: The effect of low pH on thermotolerance induction using fractionated 45°C hyperthermia. Radiat. Res. *85:* 472–479 (1981).

12 Harisiadis, L.; Sung, D., II; Kessaris, N.; Hall, E.J.: Hyperthermia and low dose-rate irradiation. Radiology *129:* 195–198 (1978).

13 Henle, K.J.: Sensitization to hyperthermia below 43°C induced in Chinese hamster ovary cells by step-down heating. J. natn. Cancer Inst. *64:* 1479–1483 (1980).

14 Henle, K.J.; Dethlefsen, L.A.: Heat fractionation and thermotolerance: a review. Cancer Res. *38:* 1843–1851 (1978).

15 Henle, K.J.; Tomasovic, S.P.; Dethlefsen, L.A.: Fractionation of combined heat and

radiation in asynchronous CHO cells. I. Effects on radiation sensitivity. Radiat. Res. *80:* 369–377 (1979).
16 Holahan, E.V.; Highfield, D.P.; Dewey, W.C.: Inducing during G_1 of heat radiosensitization in Chinese hamster ovary cells following single and fractionated heat doses. Natn. Cancer Inst. Monogr. *61:* 123–126 (1982).
17 Holahan, E.V.; Stuart, P.K.; Dewey, W.C.: Enhancement of survival of CHO cells by acidic pH after X-irradiation. Radiat. Res. *89:* 433–435 (1982).
18 Hume, S.P.; Marigold, J.C.L.: Transient, heat-induced thermal resistance in the small intestine of mouse. Radiat. Res. *82:* 526–535 (1980).
19 Hume, S.P.; Marigold, J.C.L.; Hirst, D.G.: The effect of hyperthermia on one aspect of the response of mesenteric blood vessels to radiation. Int. J. Radiat. Biol. *39:* 321–327 (1981).
20 Jähde, E.; Rajewsky, M.F.: Tumor-selective modification of cellular microenvironment in vivo: effect of glucose infusion on the pH in normal and malignant rat tissues. Cancer Res. *42:* 1505–1512 (1982).
21 Li, G.C.; Hahn, G.M.: A proposed operational model of thermotolerance based on effects of nutrients and the initial treatment temperature. Cancer Res. *40:* 4501–4508 (1980).
22 Li, G.C.; Werb, Z.: Correlation between synthesis of heat shock proteins and development of thermotolerance in Chinese hamster fibroblasts. Proc. natn. Acad. Sci. USA *79:* 3218–3222 (1982).
23 Manning, M.R.; Cetas, T.C.; Miller, R.C.; Oleson, J.R.; Connor, W.G.; Gerner, E.W.: Clinical hyperthermia: results of phase I trial employing hyperthermia alone or in combination with external beam or interstitial radiotherapy. Cancer *49:* 205–216 (1982).
24 Nielsen, O.S.: Recovery from hyperthermic damage and development of thermotolerance in unfed plateau cells in vitro. J. natn. Cancer Inst. *66:* 61–66 (1981).
25 Nielsen, O.S.: Effect of fractionated hyperthermia on hypoxic cells in vitro. Int. J. Radiat. Biol. *39:* 73–82 (1981).
26 Overgaard, J.: Effect of misonidazole and hyperthermia on the radiosensitivity of a C_3H mouse mammary carcinoma and its surrounding normal tissue. Br. J. Cancer *41:* 10–21 (1980).
27 Overgaard, J.: Fractionated radiation and hyperthermia: experimental and clinical studies. Cancer *48:* 1116–1123 (1981).
28 Overgaard, J.: Effect of hyperthermia on the hypoxic fraction in an experimental mammary carcinoma in vivo. Br. J. Radiol. *54:* 245–249 (1981).
29 Overgaard, J.; Suit, H.D.; Walker, A.M.: Multifractionated hyperthermia treatment of malignant and normal tissue in vivo. Cancer Res. *40:*2045–2050 (1980).
30 Power, J.A.; Harris, J.W.: Response of extremely hypoxic cells to hyperthermia: survival and oxygen enhancement ratios. Radiology *123:* 767–770 (1977).
31 Raaphorst, G.P.; Azzam, E.I.: Heat and radiation sensitivity of Chinese hamster V79 cells and of nine clones selected from survivors of a thermal tolerant cell population. Int. J. Radiat. Oncol. Biol. Phys. *6:* 1577–1581 (1980).
32 Song, C.W.; Kang, M.S.; Rhee, J.G.; Levitt, S.H.: The effect of hyperthermia on vascular function, pH, and cell survival. Radiology *137:* 795–803 (1980).
33 Spiro, I.J.; Denman, D.L.; Dewey, W.C.: Effect of hyperthermia on CHO DNA polymerases α and β. Radiat. Res. *89:* 134–149 (1982).

34 Spiro, I.J.; Sapareto, S.A.; Raaphorst, G.P.; Dewey, W.C.: The effect of chronic and acute heat conditioning on the development of thermal tolerance. Int. J. Radiat. Oncol. Biol. Phys. *8:* 53–58 (1982).

35 Stewart, F.A.; Denekamp, J.: Fractionation studies with combined X rays and hyperthermia in vivo. Br. J. Radiol. *53:* 346–356 (1980).

36 Stone, H.B.: Enhancement of local tumour control by misonidazole and hyperthermia. Br. J. Cancer *37:* suppl. III, pp. 178–183 (1978).

37 Thrall, D.E.; Gillette, E.L.; Dewey, W.C.: Effect of heat and ionizing radiation on normal and neoplastic tissue of the C_3H mouse. Radiat. Res. *63:* 363–377 (1975).

38 Urano, M.; Montoya, V.; Booth, A.: Effect of hyperglycemia on the thermal response of murine normal and tumor tissues. Cancer Res. *43:* 453–455 (1983).

39 Urano, M.; Rice, L.; Cunningham, M.: Effect of fractionated hyperthermia on normal and tumor tissues in an experimental animal system; in Abe, Sakamoto, Phillips, Treatment of radioresistant cancers, pp. 43–53 (Elsevier, Amsterdam 1979).

40 Urano, M.; Rice, L.; Kahn, J.; Sedlacek, R.S.: Studies on fractionated hyperthermia in experimental animal systems. I. The foot reaction after equal doses: heat resistance and repopulation. Int. J. Radiat. Oncol. Biol. Phys. *6:* 1519–1523 (1980).

W. C. Dewey, MD, Radiation Oncology Research Laboratory, CED 200, University of California, San Francisco, CA 94143 (USA)

Prog. exp. Tumor Res., vol. 28, pp. 220–231 (Karger, Basel 1984)

Feasibility of Microwave Hyperthermia for Brain Tumor Therapy

Michael Salcman

University of Maryland School of Medicine, Neuro-Oncology Service,
University of Maryland Hospital, Baltimore, Md., USA

Introduction

We have discussed the biophysical rationale for the use of hyperthermia in the treatment of malignant brain tumors previously [10, 11], but it is necessary to review some of this material briefly so we may present a fair analysis of the unique advantages and disadvantages of microwaves as a putative energy source for thermotherapy.

Heat is a physical agent: Its biologic effects are related directly to the duration and intensity of its application [4, 10]. The antineoplastic action of hyperthermia is manifest at 42 °C, but as it is a well-documented finding that metabolic damage to normal hepatocytes and neurons occurs at 43 °C [8, 14], the therapeutic index for hyperthermia is quite narrow. In the range of 44–45 °C, denaturation of protein occurs in both normal and neoplastic cells. Unfortunately, some types of naturally occurring and experimental tumors require temperatures greater than 45 °C for significant cellkilling to occur. The very nature of hyperthermia therapy, therefore, requires precisely delivered, well-controlled, and accurately measured heating capabilities [1, 2].

Clinicians have been interested in hyperthermia as an anticancer agent for more than a century, but therapeutic applications have not been developed because of the technologic problem of delivering precisely controlled and well-delimited hyperthermal fields deep within the body [11, 15]. An ideal hyperthermia system is capable of generating such fields at high temperatures without appreciably affecting normal tissues in close proximity to the tumor target. The method of heating should be versatile enough to make practical the other advantages offered by hyperthermia:

(a) repeated applications are possible because of the lack of cumulative toxicity; (b) it can be applied in concert with both chemotherapy and ionizing radiation, as hyperthermia is a powerful potentiator of these modalities [3, 6, 7], and (c) it allows precise targeting of those hypoxic and undervascularized tissues against which hyperthermia is most effective [11]. Microwaves as a physical agent and the microwave-control technology are capable of achieving many of these goals.

Microwave Interaction with Tissue

Microwaves occupy that portion of the electromagnetic energy spectrum between 300 MHz and 30 GHz – the frequency range immediately above that of radiowaves. The photon energies of microwaves are much too low to result in the ionization of biologic molecules; hence, their mechanism of action is distinctly different from that of very short-wavelength radiation, such as X-rays. When a time-varying electromagnetic field at microwave frequencies is applied to tissue, existing electric dipoles in atoms and molecules orient themselves with respect to the field, and new dipole moments are produced through the polarization of other atoms and molecules. Water is the most ubiquitous molecule, and water content is an important factor in determining the electromagnetic behavior of both normal and neoplastic tissues [5].

In comparison to other organs, the brain's gray matter is electromagnetically heterogeneous, and its dielectric properties vary with location. This variability may be ascribed to differences in local blood flow and water content [18]. Frictional movement involved in the orientation and polarization of atoms and molecules is responsible for the generation of the heat induced by microwave irradiation [2]. In addition, all tissues contain some free electrons and ions. In the presence of a microwave field, these conductive elements tend to drift until they collide with stationary atoms and molecules. The collisions provide a third mechanism for the production of heat from the interaction of microwaves with biologic materials. The magnitude of effects caused by electromagnetic interaction with tissue are frequency-dependent. This is the case within the microwave range, also.

The relative permittivity, or dielectric constant, of a substance determines whatever molecular orientation, polarization, and drift exists in that substance in the presence of a steady-state electric field at a given fre-

Table I. Microwave propagation in tissue of high water content

Frequency MHz	Wavelength in tissue, cm	Penetration depth, cm	Dielectric constant	Conductivity mho/cm
300	11.0	3.89	54	1.37
433	8.76	3.57	53	1.43
915	4.46	3.04	51	1.60
2,450	1.76	1.70	47	2.21

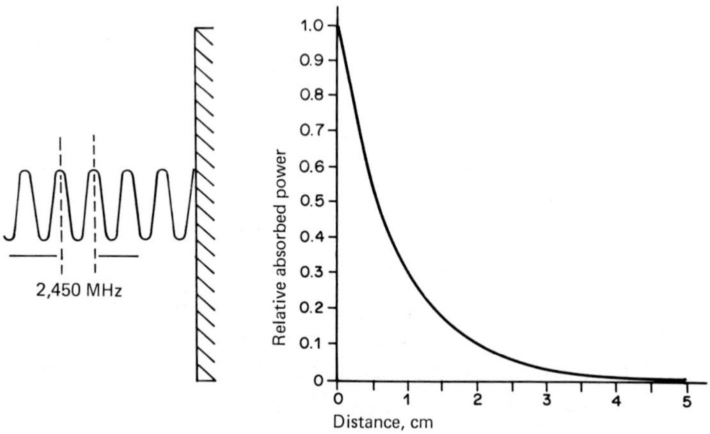

Fig. 1. Relative absorbed power versus depth in tissue. The relative absorbed power of a 2,450-MHz microwave plane wave incident at a surface is plotted against the perpendicular distance into the depth of the tissue in centimeters. Note that 50% of the power is absorbed in the first 0.5 cm, and that the 1/e point is approximately 0.85 cm.

quency. The tissue conductivity is inversely related to the dielectric constant, and both of these properties are frequency-dependent. The absorbed power per unit volume of tissue (P) is determined by the formula:

$$P = \sigma E^2, \tag{1}$$

in which σ is the conductivity and E is the magnitude of the electric field.

Tumors generally have a higher conductivity than do normal tissues; but the conductivity of gray matter is also rather high in comparison to that of other organs [2, 18]. As the propagation and attenuation of an electromagnetic wave in a material depends on the permittivity of that material, it is evident that the amount of microwave energy deposited at a given depth is a function of the frequency of the incident beam. Fortunately, the dielectric properties of the brain demonstrate relatively little variation at microwave frequencies [18]. Furthermore, the temperature coefficient of variation at microwave frequencies in brain is less than 2% per degree centigrade [5]. Nevertheless, deeper heating will occur at lower frequencies (table I). As the absorbed power density of an incident microwave beam decrements exponentially in tissue, about two-thirds of the power is absorbed at the 1/e point. The 1/e point is that point at which two-thirds of any exponential function has been dissipated:

$$A = A_0e^{-1}; (e = 2.7183) \qquad (2)$$

At the most common diathermy frequency (2,450 MHz), the 1/e point is only 0.85 cm (fig. 1). As table I indicates, further increases in the depth of penetration require progressively more drastic reductions in the microwave frequency, a situation that has profound consequences for the practical implementation of microwave hyperthermia.

The design considerations that we have reviewed so far indicate that microwaves at standard frequencies are capable of only limited penetration into tissue, that most of the power is absorbed at the surface of the organ, and that attempts to increase penetration by decreasing frequency will result in regional rather than focal hyperthermia. As the wavelength of an electromagnetic wave is inversely related to its frequency, the wavelength at 433 MHz is much longer than that at 2,450 MHz and the depth of penetration is correspondingly greater. Unfortunately, in order for the electromagnetic radiator (e.g., diathermy applicator) to be efficient, the physical size of the applicator must correspond to a significant fraction of the wavelength being emitted. Consequently, the application of microwaves at low frequencies and long wavelengths requires a treating applicator so large that it is no longer possible to restrict the focal point for heating to a small target volume. Conversely, when an applicator is physically much smaller in proportion to the wavelength of the microwaves being radiated, the fields produced in the tissue are dominated by the near-field irradiation, and very little heating is produced by the far-field irradiation. In this situation, the problem of surface heating is exacerbated, and virtually no significant amount of energy is deposited deep into the organ of interest.

In our original system design, we had hoped to use a noninvasive array of small contact radiators, applied to the skull and computer-driven in such a way that linear superimposition of the individually rapidly decaying fields would produce a hot spot within the target volume [13]. Our experiments and computer simulations soon proved this to be an unworkable approach, and we were left with the less desirable possibility of using invasive methods. Many of the physical limitations involved in producing deep heating from superficial contact applicators, however, turned out to be quite useful properties for miniature implanted devices.

Practical Systems for Brain Hyperthermia

A minimum practical system for the generation and control of a microwave hyperthermia field is illustrated in figure 2. The basic requirements consist of: a microwave source, usually a diathermy magnetrode operating at a fixed frequency, such as 2,450 MHz, or a variable source, as is shown in the block diagram; a means of amplifying and controlling the output of the generator into the useful range of 0–20 W; forward and reflected power meters for determining the actual transmitted power into the tissue; an automatic choke for cutting off the power when it exceeds some preset value; a radiator for delivering the microwaves into the tissue, and an independent means of determining the temperature achieved in the target. We have evaluated two such manual systems in our laboratories, as well as the rather more complex automated system designed by *Samaras* et al. [17].

Thermometry in a microwave field is an exceedingly complex matter. Suffice it to say that the usual metal thermistors and thermocouples cannot be used because the electric field set up by the microwaves will cause current to flow in conventional temperature probes, and self-heating or apparent local hot spots will result. Therefore, nonperturbing probes must be used; these usually are fabricated from nonconducting optical fibers connected to either liquid crystal, semiconductor, or fluorescent optical sensors [2]. We have used successfully both the Luxtron fluoroptic sensor and the Vitek high-resistivity thermistor sensor to map thermal fields in surrounding brain, as well as in the periphery of the brain tumor. These probes have an outer diameter of 0.8 and 1.2 mm, respectively, and consequently must be handled with considerable care. Central temperatures in the field can also be obtained by placing one of these probes near the

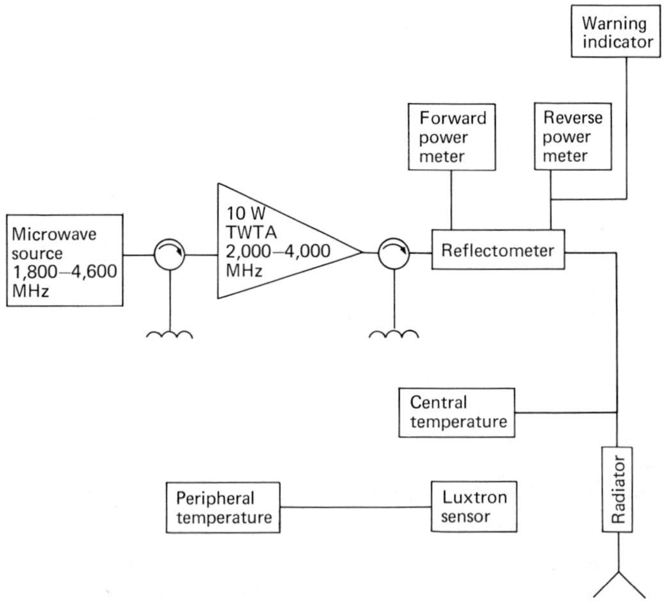

Fig. 2. Block diagram of a manual-control microwave hyperthermia system. The output of a microwave source is fed to a traveling-wave tube amplifier (TWTA) capable of boosting power over a specified frequency range. The actual power going into the tissue (the transmitted power) equals the difference between the readings on the forward and reflected power meters; the measurements are made in a reflectometer. When the reflected power is too high, a warning light goes on and the system shuts down. Central temperature in the field is measured from a thermistor bonded to the radiating antenna; temperatures elsewhere in the field are measured by a nonperturbing Luxtron probe. Feedback control on the output of the TWTA is provided by visual observation of the central temperature.

radiator or by bonding a high-impedance 1-MΩ thermistor (0.014£) directly to the antenna and using appropriate corrections on the readings [17].

Technical Feasibility

The radiator depicted in figure 3a has undergone extensive laboratory and clinical testing [19]. In the normal cat brain, the miniature microwave antenna produces an ellipsoidal thermal field that is centered at the point where the outer insulator ends. At 2,450 MHz, this field is approximately 2 cm in diameter, and a central temperature of 45°C is achieved at less

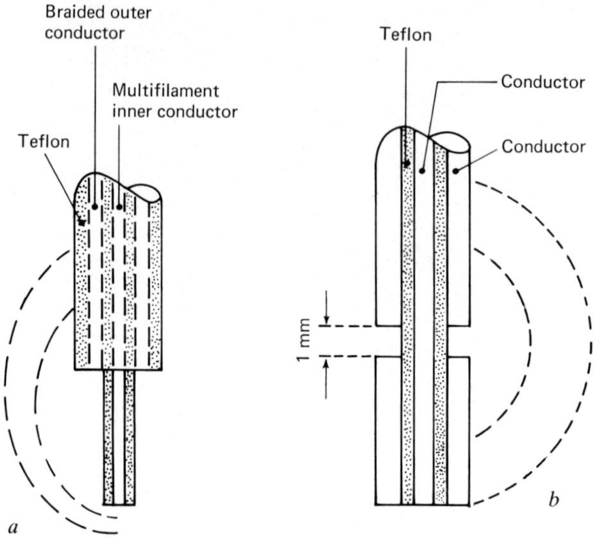

Fig. 3. Interstitial microwave irradiators. The coaxial cable *(a)* radiates an ellipsoidal field centered at the point where the needle antenna emerges from the cable; it is flexible and has an outer diameter of 1.8 mm. The slot antenna *(b)* is semirigid and made of tubing with an outer diameter of 1 mm or less; the field is centered around the slot and is roughly spherical.

than 10 W steady state [10, 11]. Less than 60 s are required to move from an initial temperature of 37 to 45°C near the antenna. When power is cut off, rapid cooling occurs, mainly because of the high blood flow in the normal brain, and the brain returns to a baseline temperature in less than 60 s. No abnormalities in arterial pressure, heart rate, the electrocardiogram, or pupillary function are observed during or after 60 min of interstitial heating at the peak temperature; this is in contrast to the many abnormalities, including herniation, observed during and after 60 min of whole-body heating at 43°C in the cat [14]. Measurements made using a subarachnoid bolt indicate that in the normal cat brain, focal interstitial hyperthermia does not produce any elevation of the intracranial pressure.

Clinical Feasibility

The same radiator (fig. 3a) has been tested in an initial series of 6 patients. The detailed results of this phase-I trial are presented elsewhere

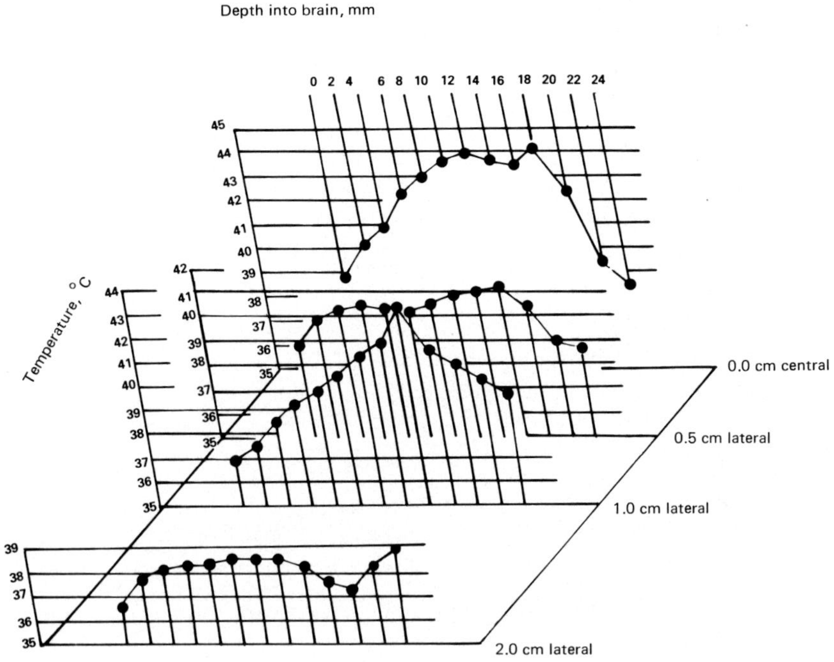

Fig. 4. Thermal-field plot of a microwave radiator in a human brain tumor. Three-dimensional plot of temperatures recorded on the y axis at various longitudinal distances along the antenna (x axis) as it runs from the surface of the brain into the depths, and similar measurements made along other perpendicular penetrations of a nonperturbing thermistor probe at various radial distances from the antenna (z axis). Note that with a peak temperature of 44°C at the center of the field, brain temperatures remain mildly elevated at radial distances of 2 cm from the long axis of the antenna [from ref. 11, with permission].

[12]. One of the main objectives of this phase-I study was to obtain biophysical measurements, in keeping with our philosophy that feasibility trials of a clinical device should mimic, as nearly as possible, the experimental and data-collection procedures of device-testing in animals [9].

The miniature microwave antenna was implanted in patients with recurrent primary brain tumors under direct vision during open surgery. Usually, it was implanted in a nonresectable portion of the mass after significant mechanical removal had been performed. Intraoperative heating and cooling curves were obtained and thermal field plots were constructed; about 20 min were required to obtain these data. A high-resistance thermistor bonded to the antenna served as the source of central

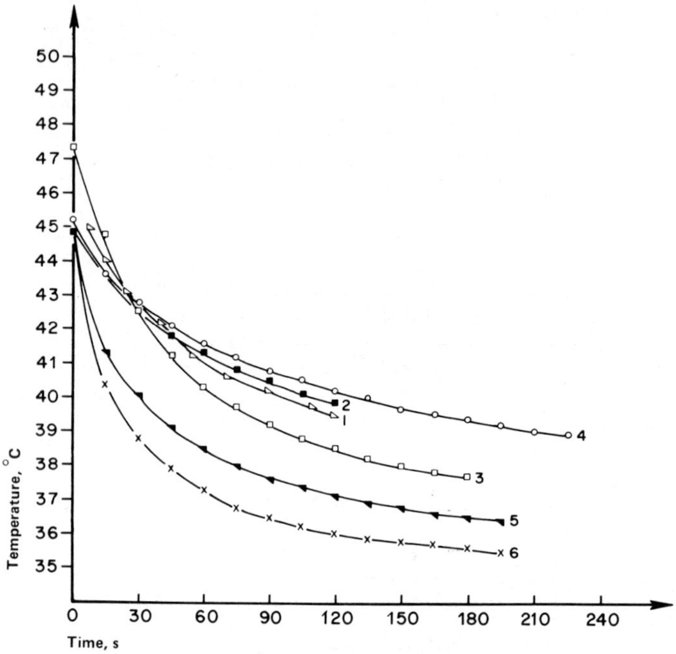

Fig. 5. Intraoperative cooling curves from 6 patients. The temperature recorded from the center of the field is plotted against time in seconds for each of the first 6 patients undergoing implantation. Note that most tumors did not return to baseline temperatures until 120–180 s had passed. Data were obtained manually from the first 3 patients, and by computer from the last 3 patients [from ref. 12, with permission].

temperatures, and a hand-held Vitek or Luxtron nonperturbing probe was used to scan the cortex at various depths from the surface and at several radial distances from the central antenna. It was found that the thermal field in a human brain tumor was approximately 4 cm in diameter when operated at 2,450 MHz and at a central temperature of 45°C [16] (fig. 4). Similarly, cooling curves at the antenna were much slower in the brain tumor than they had been in the normal cat brain (fig. 5). In some patients, it took more than 180 s to reach baseline temperatures. It thus was clear that human brain tumors did not dissipate heat efficiently, presumably because of ineffective or depressed blood flow.

At the conclusion of surgery, the transcutaneous cable connection of the antenna had to be threaded laboriously through colinear apertures in the dura, bone, and scalp. A subarachnoid bolt was implanted over the contralateral frontal lobe in all 6 patients as a safety precaution, and in order to correlate intracranial pressure data with that obtained previously in animals. An immediate postoperative computerized tomography (CT) scan confirmed the intended position of the device in 5 of the 6 patients; in our 2nd patient, the antenna had moved after the application of the head-dressing and its position was readjusted under CT control. The artifact produced by the antenna was obvious but not intolerable. The evening of surgery, all patients received a 60-min session of hyperthermia at 45 °C; this was repeated 48 h after the operation, and the antenna was withdrawn immediately afterward. 1 of the 6 patients received 1,3-bis(2-chloroethyl)-1-nitrosourea (BCNU) intravenously, simultaneously with her heat treatments and on the intervening day. No patient was physically aware of either the onset or the cessation of treatment.

The first 3 patients were treated with a simple manual system and the second 3 with the more complex automated-control system. Although the manual unit required constant vigilance on the part of the operator to provide just enough power to keep the temperature constant, it was easy to use and performed satisfactorily. The automated system provided its own feedback control, bookkeeping, and error-checking. At the end of each session, it was possible to produce all of the relevant biophysical and treatment data immediately in graphic form [12].

Only one hyperthermia session produced a transient neurologic deficit – an increase in a preexisting dysphasia that resolved within 2 h after the end of treatment. No patient had an elevation of intracranial pressure greater than 1 mm Hg within the normal range (0–15 mm Hg), and there were no permanent neurologic sequelae or delayed deaths that could be attributed to the hyperthermia sessions. 2 patients developed a cerebrospinal-fluid leak along the track of the antenna – one because of a technical error, in that we had neglected to use gelfoam to pack the space around the antenna as it pierced the dura. Both patients required surgical treatment for their respective subdural and subgaleal fluid collections.

Prospects and Limitations

At this time, we really do not know if the survival of patients with malignant astrocytoma will be improved through the simultaneous use of hyperthermia, radiation, and chemotherapy [10]. We are presently initiating a phase-I/II trial in which interstitial hyperthermia and interstitial irradiation will be combined in newly diagnosed cases of glioblastoma multiforme.

It is clear that our present techniques for inducing interstitial hyperthermia with microwaves are both safe and feasible. Furthermore, anecdotal evidence from 2 of our 6 patients who have survived 1 and 2 years, respectively, from the time of tumor recurrence and who have negative CT scans at their latest follow-up indicates that long-term survival can be obtained with hyperthermia as postoperative therapy. In addition, 1 of our patients demonstrated excellent clinical tolerance and response to same-day use of radical surgery, chemotherapy, and hyperthermia. Since the greatest potential benefit of hyperthermia may well be in treating those tumors that are surgically inaccessible, it is important that we develop techniques for stereotactically placing microwave antennas in deep brain structures. The new antenna designed by *Samaras* (fig. 3b) is semirigid and has been fitted to the CT-compatible Leksell stereotactic system. It is our hope that this configuration will obviate the many difficulties involved in prolonged transcutaneous connections. Finally, it is important to remember that no hyperthermia system is capable of delivering a spatially homogeneous field, and that we must remain vigilant concerning the possible future complications of combination therapies involving this modality.

Acknowledgment

This work was supported in part by Public Health Service Grant 1-R01-NS16316 from the National Institute for Neurological and Communicative Disorders and Stroke, American Cancer Society Grant PDT-105, and the Neuro-Oncology Research Fund at the University of Maryland.

I am especially grateful for the close collaboration of my engineering colleagues, Dr. *George M. Samaras* and Mr. *Gordon Neuberth*.

References

1 Bowman, H.F.: Heat transfer and thermal dosimetry. J. Microwave Power *16:* 121–133 (1981).
2 Christensen, D.A.; Durney, C.H.: Hyperthermia production for cancer therapy: a review of fundamentals and methods. J. Microwave Power *16:* 89–105 (1981).
3 Crile, G., Jr.: The effects of heat and radiation on cancers implanted on the feet of mice. Cancer Res. *23:* 372–380 (1963).

4 Dickson, J.A.; Calderwood, S.K.: Temperature range and selective sensitivity of tumors to hyperthermia: a critical review. Ann. N.Y. Acad. Sci. *335:* 180–205 (1980).

5 Foster, K.R.; Schepps, J.L.: Dielectric properties of tumor and normal tissues at radio through microwave frequencies. J. Microwave Power *16:* 107–119 (1981).

6 Hahn, G.M.: Potential for therapy of drugs and hyperthermia. Cancer Res. *39:* 2264–2268 (1979).

7 Li, G.C.; Evans, R.G.; Hahn, G.M.: Modification and inhibition of repair of potentially lethal X-ray damage by hyperthermia. Radiat. Res. *67:* 491–501 (1976).

8 Pettigrew, R.T.; Galt, J.M.; Ludgate, C.M.; Horn, D.B.; Smith, A.N.: Circulatory and biochemical effects of wholebody hyperthermia. Br. J. Surg. *61:* 727–730 (1974).

9 Salcman, M.: Clinical hyperthermia trials: design principles and practice. J. Microwave Power *16:* 171–177 (1981).

10 Salcman, M.; Kaplan, R.S.; Samaras, G.M.; Ducker, T.B.; Broadwell, R.D.: Aggressive multimodality therapy based on a multicompartmental model of glioblastoma. Surgery *92:* 250–259 (1982).

11 Salcman, M.; Samaras, G.M.: Hyperthermia for brain tumors: biophysical rationale. Neurosurgery *9:* 327–335 (1981).

12 Salcman, M.; Samaras, G.M.: Interstitial microwave hyperthermia for brain tumors: results of a phase-1 clinical trial. J. Neuro-Oncol (in press, 1983).

13 Salcman, M.; Samaras, G.M.: Neurosurgery and clinical engineering. J. clin. Eng. *3:* 251–256 (1978).

14 Salcman, M.; Samaras, G.M.; Mena, H.; Monteiro, P.; Garcia, J.: Wholebody hyperthermia: potential hazards in its application to glioblastoma; in Paoletti, Walker, Butti, Knerich, Multidisciplinary aspects of brain tumor therapy, pp. 351–356 (Elsevier, Amsterdam 1979).

15 Samaras, G.M.; Cheung, A.Y.: Microwave hyperthermia for cancer therapy. CRC crit. Rev. Bioeng. *5:* 123–184 (1981).

16 Samaras, G.M.; Salcman, M.; Cheung, A.Y.; Abdo, H.S.; Scheep, R.S.: Microwave-induced hyperthermia: an experimental adjunct for brain tumor therapy. Natn. Cancer Inst. Monogr. *61:* 477–482 (1982).

17 Samaras, G.M.; Van Horn, H.W.; King, V.F.; Slawson, E.L.; Cheung, A.Y.: Clinical hyperthermia systems engineering. J. Microwave Power *16:* 161–169 (1981).

18 Stuchly, M.A.; Athey, T.W.; Stuchly, S.S.; Samaras, G.M.; Taylor, G.: Dielectric properties of animal tissues in vivo at frequencies 10 MHz – 1 GHz. Bioelectromagnetics *2:* 93–103 (1981).

19 Taylor, L.S.; Samaras, G.M.; Cheung, A.Y.; Salcman, M.; Scott, R.M.: Implantable microwave radiators for clinical hyperthermia. Radio Science *17:* 1255S–1335S (1982).

M. Salcman, MD, Associate Professor of Neurological Surgery, University of Maryland School of Medicine, and Chief of the Neuro-Oncology Service, University of Maryland Hospital, Baltimore, MD 21201 (USA)

Prog. exp. Tumor Res., vol. 28, pp. 232–245 (Karger, Basel 1984)

Feasibility of Treating Malignant Brain Tumors with Focused Ultrasound

Richard H. Britt, Douglas W. Pounds, Bernard E. Lyons

Division of Neurosurgery, Stanford University School of Medicine, Stanford, Calif., USA

Introduction

Hyperthermia, either by itself or in combination with radiation therapy and chemotherapy, holds promise as a new modality for the treatment of malignant brain tumors. Primary brain tumors are ideally suited to localized hyperthermia therapy for 3 reasons: (1) these tumors are usually single lesions and rarely metastasize; (2) the 3-dimensional volume and shape of the tumor relative to the normal brain can be determined by a variety of imaging techniques (computerized tomography (CT), nuclear magnetic resonance, and high-resolution B-mode ultrasound scanning), and (3) the brain and tumor tissue are fairly homogeneous in density in comparison with other body cavities such as the abdomen or thorax, where air, tissue, and bone interfaces exist.

The concept of using ultrasound to treat malignant cancer dates back nearly 50 years [14]. Renewed interest in using hyperthermia for cancer treatment has evolved from encouraging tissue culture and animal studies and from new and improved technology for generating localized heat in tissue.

Induction of Hyperthermia Using Ultrasound in Brain

In ultrasonic hyperthermia systems, high-intensity sound waves are used to generate heat in tissues. Acoustical waves deposit energy when they flow through tissue. The rate of energy transfer is determined by the intensity of the wave, by its frequency, and by the mechanical characteristics of the tissue (absorption coefficient). The attenuation of ultrasonic energy through tissue is exponential:

$$I = I_0 e^{(-2\mu x)},$$

where I is the acoustical intensity at depth 'X' in tissue (W/cm^2) and μ is the attenuation coefficient for a particular frequency and tissue (Np/cm). This means that in a tissue of homogeneous density, the temperature is highest at the surface and decays exponentially relative to depth. Surface cooling can modify the temperature pattern by thermal conduction, reducing the heat produced at the site of maximum ultrasound field intensity.

The absorption of ultrasonic energy in tissue increases almost linearly relative to frequency throughout most of the range commonly used in hyperthermia therapy (0.3–6 MHz). The operating frequency is largely determined by the particular application; for example, at lower frequencies, less of the ultrasonic energy is absorbed over a specified path length, allowing more of the original energy to be delivered to the deeper tissues. However, the delivered energy is less efficient at heating tissue, and more residual energy remains beyond the target region, potentially to cause undesired heating in other tissues. These considerations were taken into account in designing the ultrasound hyperthermia treatment systems that we discuss.

A block diagram of the ultrasound hyperthermia treatment system developed for our experimental brain tumor studies is illustrated in figure 1. Ultrasonic fields are generated by a transducer containing a 4-cm, 1- or 2-MHz, lead zirconium titanate (PZT) piezoelectric crystal (fig. 2). This unfocused single-beam ultrasound transducer has 2 basic compartments: (1) the piezoelectric crystal with interfacing electronics, and (2) a plexiglass housing through which temperature-controlled, degassed water circulates (fig. 2). A frequency synthesizer generates a continuous sine wave at the resonance frequency of the piezoelectric crystal that is amplified by a RF power amplifier. A power meter measures the relative efficiency of the conversion of electrical energy to mechanical energy (sound waves) by measuring the power forward versus the power reflected. The typical unmatched electrical-to-acoustical conversion efficiency of a PZT-based ultrasound transducer at resonance is about 40%. The sound waves produced by the piezoelectric crystal have to be effectively coupled to the tissue. To accomplish this, a plexiglass housing provides a water path from the piezoelectric crystal to the exit membrane. Degassed water circulates through this housing, which is necessary to prevent the formation of small

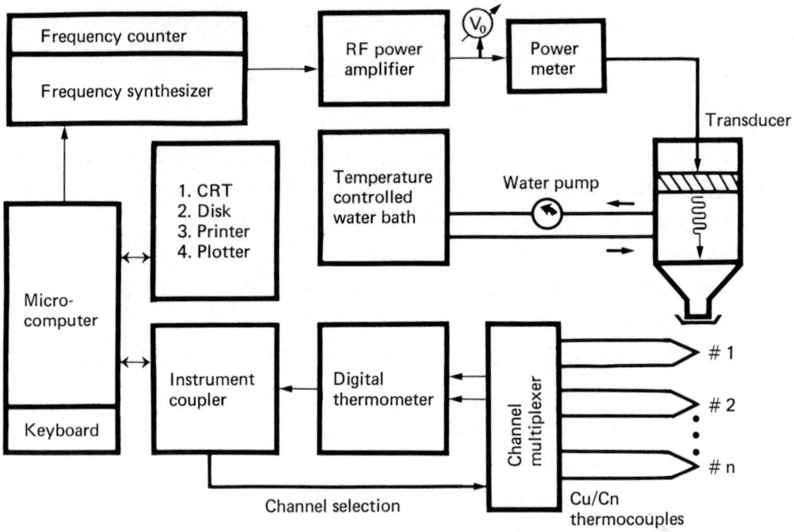

Fig. 1. Block diagram of the experimental hyperthermia treatment system.

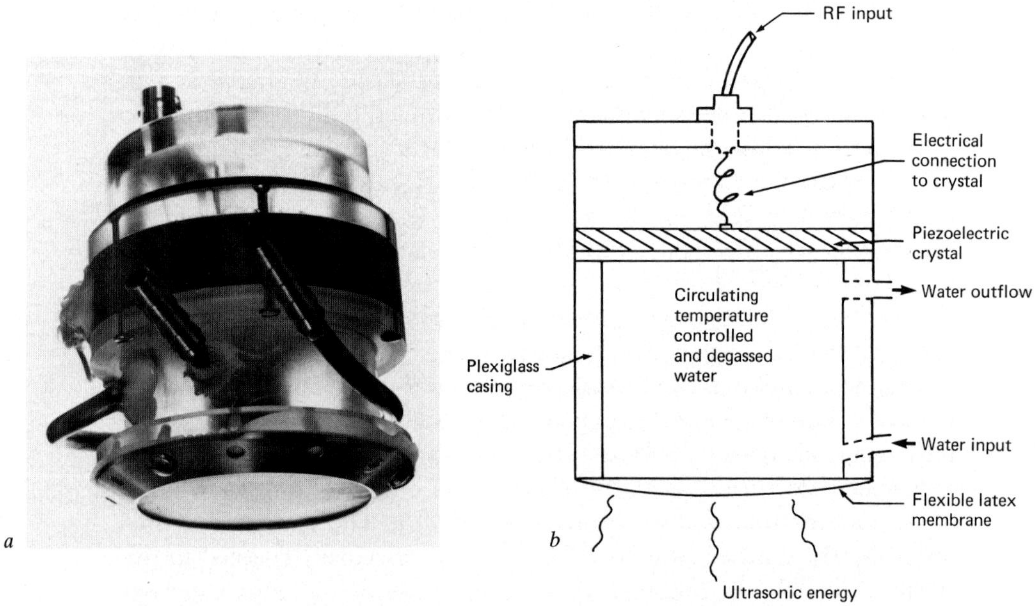

Fig. 2. a The 2-MHz ultrasound transducer used for treating brain tumors in dogs. *b* Schematic diagram of a nonfocused ultrasound transducer.

air bubbles that would scatter the ultrasonic beam. A flexible membrane or cuff on the end of the housing allows for good tissue coupling through either an ultrasound gel or saline. The circulating water temperature may be varied to provide temperature control of the near surface tissues. In addition, since phase interference patterns in near proximity to the resonating crystals may generate wide variations in field intensity, the length of the water housing may be selected for an optimum intensity distribution.

Effects of Ultrasound Hyperthermia on Normal Brain

Before ultrasound hyperthermia can be used to treat brain tumors effectively, the tolerance of the normal brain to heat must be thoroughly understood. The effects of ultrasound in brain tissue can be viewed in terms of changes in structure and function. In either case, one must further consider whether the changes are reversible or permanent.

Irreversible structural changes (i.e., lesion production) using ultrasound in the brain have been studied by several groups of investigators [1, 8, 13, 17]: cat, rat, monkey, and human brains appear to respond identically [13]. This conclusion is of fundamental importance if animal tests are to be used to establish the parameters for initial clinical use. Little research defining ultrasonic effects in brain tissue has been correlated with temperature distributions. Workers in our laboratory have been investigating the threshold for damage as a function of ultrasound intensity and resulting temperature distribution in a series of acute experiments in normal cat brain [3, 4]. Detailed thermal distributions were determined by using an array of three 26-gauge, triple-junction thermocouple probes. The probes left a track that was readily identified histologically. Serial coronal brain sections were cut and stained, allowing for precise correlation of histology, thermocouple location, and temperature.

At a temperature of 42°C, no evidence of acute damage could be detected in either the gray or white matter for heating periods of up to 50 min (fig. 3). At a temperature of 42.5°C, neurons developed a pyknotic appearance compatible with irreversible structural damage. At 43°C, there was partial loss of neurons in the gray matter (fig. 3). At the same temperature, cerebral edema separated the tracts within the white matter (fig. 3). At temperatures of 44–45°C, there was loss of neurons in both cortical and subcortical gray matter (fig. 4).

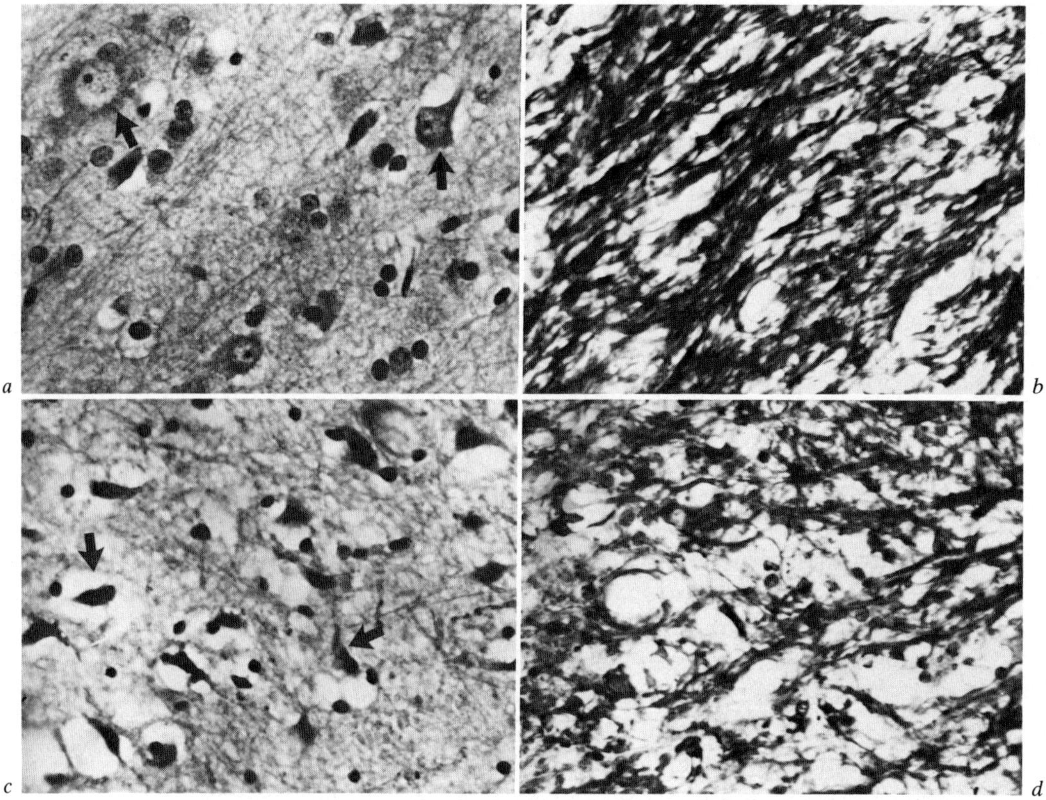

Fig. 3. Histological findings in cat brain resulting from ultrasound heating at subthreshold levels for tissue damage. *a* Normal-appearing cortical neurons (arrows) near a thermocouple sensor at 42°C. *b* Normal-appearing white matter at 42°C. *c* At a temperature of 43.4°C, the cortical neurons appear pyknotic, with loss of normal cellular structures (arrows). *d* Heating white matter to 43.8°C results in edema separating the myelinated nerve fibers, but no lesions are noted (magnification 480×, before reduction).

At temperatures greater than 45°C, sharply marginated, ovoid or spheroidal lesions were created (fig. 4). Photomicrographs of the hippocampus showed that the boundary between normal-appearing neurons and areas of complete lysis or disappearance of neurons was only a few hundred microns (fig. 4). This narrow transition between healthy and damaged brain was reflected in a rapid falling off in temperature as a function of distance from the ultrasound field.

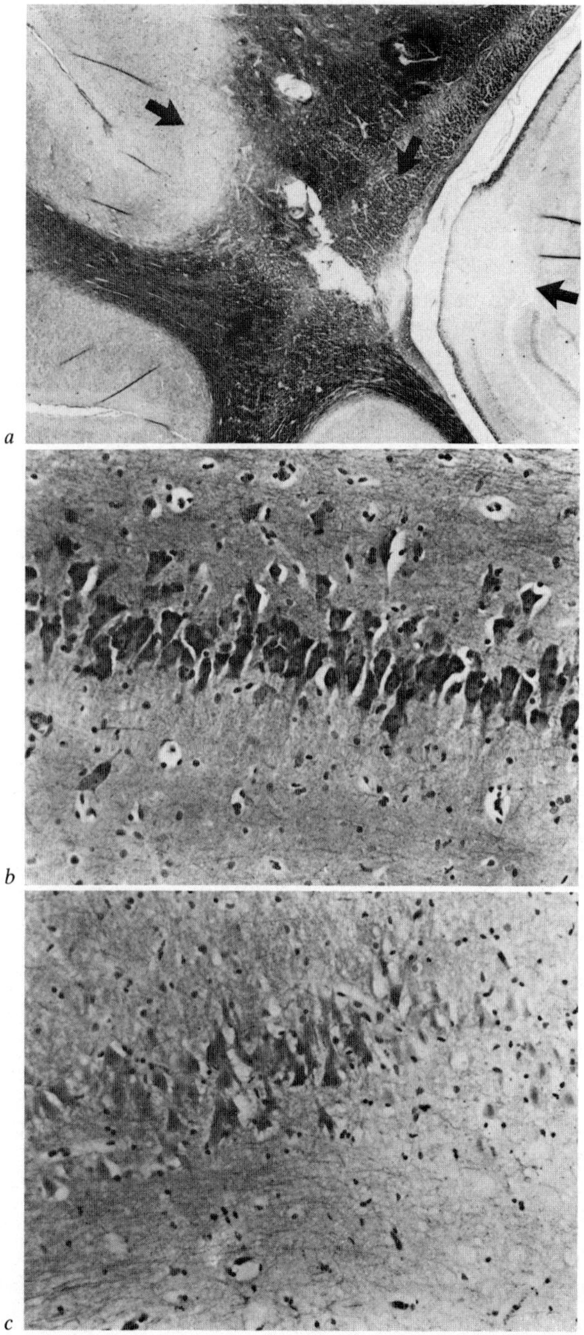

Fig. 4. *a* An oval lesion (surrounded by black arrows) was created by ultrasound in the deep white matter and hippocampus: average deposited power was 0.4 W/cm^3; highest recorded temperature was 47°C; length of heating was 50 min; cuff temperature was 37°C (magnification 14× before reduction). The lesion in the hippocampus shows that, over a distance of 500 μm, the neurons go from healthy-appearing cells *(b)* to a transition zone *(c)* of abnormal pyknotic cells and lysed neurons (magnification 312× before reduction).

There are few studies assessing the neurophysiologic state of the brain in relation to temperature. Using regional perfusion of the dog brain, *Harris* et al. [12] found that temperatures of up to 42°C were tolerated without producing changes in the electroencephalogram (EEG) or visual-evoked response. Above 42°C, there was a significant slowing in the EEG. The visual-evoked response lost its negative peak between 37 and 42°C, and there was a mild reduction in the amplitude of the positive peak. If the brain temperature was raised above 43°C, the response rapidly deteriorated and then disappeared entirely within 10 min. Dogs in which temperatures were maintained at 42°C survived without developing neurologic deficit; however, those exposed to temperatures above 44°C developed hemiparesis, anisocoria, and frequently did not regain consciousness after the heating procedure. The histologic analyses in this study correlated with postoperative status. In dogs in which the heating was applied at temperatures of 42°C or less, no evidence of cytologic change was detected in the brain. For those heated to temperatures over 44°C, there was evidence of infarction, focal hemorrhage, and brain edema. The important conclusion from this work is that permanent functional changes will not occur unless a lesion is produced. Workers in our laboratory are currently investigating changes in neural functions that occur with heating in the temperature range of 40–45°C using a whole-body, large-animal hyperthermia model. Changes in multimodality evoked potentials (auditory brainstem, visual, and somatosensory evoked potentials) are correlated with structural changes in the spinal cord, brainstem, and the subcortical and cortical areas that generate these responses. In addition, the cytoarchitecture of the areas of the brain known to be most susceptible to metabolic alterations (e.g., cerebellum and hippocampus) are studied for evidence of thermal damage.

Dubois et al. studied the EEG [6] and somatosensory evoked responses [7] as a function of temperature in humans undergoing whole-body hyperthermia treatments for cancer. During heating, the EEG recordings showed major changes in both rhythmicity (slower) and amplitude (lower), indicating a diffuse decrease of cortical activity. These changes were reversed by cooling. Somatosensory evoked potentials significantly decreased in amplitude over the 2-hour period of heating and were totally absent during the plateau phase of the treatment (41.8–42°C). However, with cooling, the response returned to normal.

Of considerable clinical importance is that radiation may modify the temperature at which structural and functional effects occur in the central nervous system (CNS) [10]. Of some concern is the recent report of 3 cases in which patients receiving whole-body hyperthermia (41.8 °C) had sudden myelopathy following spinal cord irradiation [5]. Because hyperthermia will be used in conjunction with ionizing radiation to treat brain tumors, further work must be done in animal models to determine the relative dose of radiation that can be effectively combined with hyperthermia without causing radiation necrosis.

From these structural and functional studies, it appears that 42 °C is the threshold above which irreversible damage is produced in the CNS. Tissue culture studies of human glioblastoma cells indicate that only after 300 min of heating at 42 °C does there begin to be a marginally lethal effect on these malignant cells, with cell survival rates of 70–80% [9]. At 43 °C, only 10% of the cells survived after 300 min of heating. Above 43 °C, cell survival rates decreased more rapidly. Although in vitro studies do not always predict results obtained in vivo, the possibility exists that effecting damage to malignant tissue may require temperatures above the threshold for damage to normal brain. This suggests that the precise localization of energy to the region of malignant cells may be required.

Application of Ultrasound Hyperthermia to the Treatment of Brain Tumors

The problem in applying hyperthermia techniques to brain tumor treatment is delivering the required power to the region of the brain infiltrated with tumor without overheating the surrounding normal brain. This problem is complicated by large variations in the extent of blood flow within each histologic type of brain tumor. Developed tumors typically have a necrotic center of low perfusion surrounded by a well-perfused perimeter. The predominant factor determining how much energy is required to heat a particular tissue is the rate at which blood flow carries heat away. The high rate of blood perfusion in the brain (50–60 ml/100 g/min) means that significant power deposition must occur for this tissue to heat. It requires about 0.04 W/cm^3 of deposited power in brain tissue to elevate the temperature 1 °C. In order to heat a brain tumor with normal blood flow to 43 °C, a deposited power in excess of 0.2 W/cm^3 is required. If the brain tumor has a perfusion greater than the surrounding brain, greater power deposition will be required to heat the tumor tissue, thereby increasing the risk of overheating the surrounding brain if the energy is not precisely deposited.

Using a single transducer system, heating can be accomplished to a depth of 2–3 cm, which is adequate for experimental brain tumor research in animals. As a specified unfocused ultrasound transducer has a fixed beam width, a variety of transducers may be required to treat tumors of different diameters. We are using the avian Rous sarcoma virally-induced brain tumor model in dogs [2]. If the virus is deposited in the white matter in newborn puppies, anaplastic gliomas develop within 4–6 weeks. Tumor growth can be monitored using both CT and high resolution ultrasound scanning techniques. These same imaging techniques will be used to set treatment fields and to monitor the effects of treatment. The first generation of ultrasound transducers built in our laboratory have beam widths of from 2 to 4 cm, corresponding to the typical range of tumor diameters in this model.

There are limitations to using a single unfocused ultrasound transducer for tumor heating. With a uniform ultrasound field, temperature rise will be inversely proportional to perfusion. In regions of low perfusion in the central necrotic area, the temperature will rise to much higher levels than it will in the well-perfused periphery. In addition, for tumors of irregular geometry, a significant amount of normal brain will have to be included in the ultrasound field. Despite these limitations, it provides a relatively simple system to begin investigating the effects of ultrasound hyperthermia on a brain tumor model that has clinical and histologic features resembling the human glioblastoma.

The treatment of human brain tumors will require an ultrasound hyperthermia treatment system capable of heating large volumes of tumor to depths that may exceed 7–10 cm. Using ultrasonic energy to heat tissues at these depths requires either a focused single transducer system [16] or superimposition of several ultrasound beams using different portals of entry [11]. The wavelengths at ultrasonic frequencies (1.5 mm at 1 MHz in water) allow the energy to be focused by techniques similar to those used in focusing light. Well-defined beam geometries can therefore be generated to heat small, discrete volumes either at the surface or in deep tissue. The details of designing a focused ultrasound transducer are discussed by *Lele* [16]. In order for a volume of tissue to be heated, the focused transducer must be moved. *Lele* [15] has designed a system that is capable of heating larger volumes of tissue in experimental animals at depths of 1–2 cm. A single movable focused ultrasound transducer that is capable of heating a large volume of deep-seated tumor without overheating the intervening normal tissue has yet to be developed.

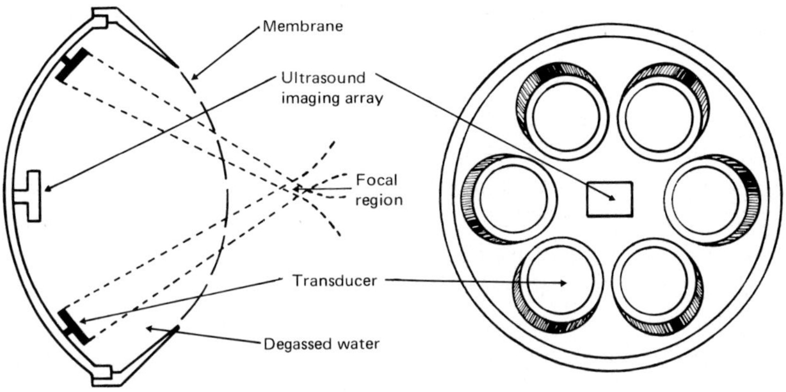

Fig. 5. Schematic drawing of the Stanford isospherical system.

Another approach for heating large, deep-seated tumors using overlapping ultrasonic beams was developed by *Pounds* and colleagues at Stanford University [11]. This isospherical, ultrasonic hyperthermia treatment system consists of 6 transducers placed in a spherical housing containing temperature-controlled, circulating, degassed water (fig. 5). The system uses 6.5 cm diameter, 0.36 MHz, plane disk transducers that are mounted so that their beams may superimpose or approach in close proximity at a point approximately 25 cm distant from the face of the transducers. A flexible membrane forms the anterior surface of the housing and makes the interfacing contact with the patient. The focal region of heating is centered for the proper depth by adjusting the water pressure in the housing to distend the flexible membrane to the appropriate distance. With this system, the maximum working distance between the membrane and the center of the heated volume is about 15 cm. Tumors up to that depth have been heated without inducing excessive temperatures in the surrounding tissues. Experimental studies heating normally perfused pig-thigh muscle showed that the maximum volume of tissue that can be heated is an ellipsoid with respective axes of about 6 and 7 cm. The second generation of isospherical system currently being developed, may incorporate an ultrasound imaging array that could be used for tumor localization and for tracking the progress of treatment.

In general, a problem with all heating systems that has become evident during the application and monitoring of hyperthermia in various tumors, in normal tissues, and in different locations is that uniform power exposure or deposition generally does not produce uniform temperatures because of variations in perfusion. The blood flow at various locations in a tumor is, for the most part, unpredictable and uncontrollable, thereby making temperature control throughout a tumor difficult, if not impossible, to achieve with present techniques.

An approach that offers promise for circumventing this problem of variable perfusion is being developed and tested in the Division of Neurosurgery at Stanford University. The approach is based on the time-temperature relationship that is recognized in conditions of hyperthermia, and predicts that an equivalent hyperthermic effect can be reproduced if a change in treatment temperature is compensated by an inverse change in the duration of treatment time. Lesion production in the brain as a function of time and temperature was investigated by *Lele* [17]. The temperature for threshold damage varied from 66.5 ± 2.5 °C sustained for a duration of 0.3 s, down to 43 ± 0.5 °C for a duration of 900 s. *Lele* [17] found that the elevation of temperature to 42.5 °C did not result in morphologically detectable damage, even when sustained for periods of up to 8 h. This level is within 0.5 °C of the threshold determined in our laboratory for heating periods of 50 min [3, 4].

In the technique that we call 'rapid hyperthermia', a high-intensity region of ultrasonic energy is used to elevate temperature sufficiently in a small volume of tissue. The period of heating is adequately brief that sufficient blood flow into the heating volume does not occur. By this technique, an unknown parameter – perfusion – is exchanged for parameters that can be determined experimentally (i.e., time-temperature curve for damage threshold, tissue absorption, and exposure power).

The 3-dimensional matrix of tumor will be defined by CT or other scanning techniques, such as nuclear magnetic resonance imaging or high-resolution ultrasound scanning, and will be coupled directly to the treatment-system computer. The ultrasound system will consist of multiple transducers with the assembly attached to a stereotaxic frame. Using a series of computer-controlled stepping motors and rotational assembly on the stereotaxic frame to direct the tightly focused high-intensity field of ultrasonic energy, the entire volume of brain tumor can be scanned sequentially.

Such a system will be able to treat large, geometrically irregular volumes of brain tumor. The critical time-temperature threshold for damage to normal brain tissues will be maintained whenever the interface between healthy and diseased tissues is sonicated. However, the high-intensity scanning field will be considerably smaller than the tumor, to allow for sterilization of the malignant cells whenever the scanning field is safely within the tumor volume. This system, when developed, will be tested on both induced and spontaneous brain tumors in dogs before proceeding to clinical trials in humans.

To establish a thermal treatment plan for brain tumor hyperthermia treatment, research will have to be done in both normal brain tissue (tissue toxicity studies) and in experimental tumor models. Before adequate planning of hyperthermia therapy can be formulated, the following information must be determined: (1) the optimal sequencing of hyperthermia treatments as well as the parameters of treatment; (2) whether hyperthermia treatments should be given before, after, or during the application of ionizing radiation; (3) whether thermal tolerance develops in both normal brain and tumor tissue; (4) if thermal tolerance decays more quickly in tumor tissue, allowing higher treatment temperatures at the normal brain-tumor interface, and (5) the interaction of hyperthermia with ionizing radiation and chemotherapeutic drugs in normal brain, to avoid damage in healthy tissue.

Thermometry

Temperature monitoring and detailed mapping of the thermal distribution are essential parts of any hyperthermia research program. At the present time, temperature monitoring requires the invasive placement of fine-gauge probes. For ultrasound studies, fine-gauge thermocouple needle probes made of copper and constantan are generally used. A computer-controlled thermometry system has been designed in our laboratory [4]. If temperature measurements exceed the desired maximum temperature, a feedback circuit can be activated to reduce the ultrasound intensity by reducing the amplitude of the sine wave produced by the frequency synthesizer (fig. 2).

Conclusion

Ultrasound used in rapid hyperthermia techniques is uniquely suited to the treatment of brain tumors because therapeutic temperatures can be precisely limited to the defined volume of tumor, sparing the surrounding normal brain. The system can deliver adequate power to the required depths within the brain. Although a bone window is required for the procedure, the bone flap may be stored in a frozen state and replaced during a subsequent operation, thereby allowing multiple treatments over a period of weeks or months, when necessary. The bone window will also permit the use of ultrasonic imaging to monitor each treatment and follow tumor regression.

Acknowledgements

This research is supported by PHS Grant CA29731 awarded by the National Cancer Institute, an Institutional American Cancer Society Award and the Appert Memorial Fund. The authors thank *Elizabeth Millerman* for histology; *Peter Richey* for graphics; *Peter Richey, Helene Englander,* and *Phil Horne* for photography; and *Laura Ziock* for manuscript preparation.

References

1 Barnard, J.W.; Fry, W.J.; Fry, F.J.; Brennan, J.F.: Small localized ultrasound lesions in the white and gray matter of the cat brain. Archs Neurol. Psychiat. *75:* 15–35 (1956).

2 Bigner, D.D.; Kvedar, J.P.; Shaffer, T.C.; Vick, N.A.; Engel, W.K.; Day, E.D.: Factors influencing the cell type of brain tumors induced in dogs by Schmidt-Ruppin Rous sarcoma virus. J. Neuropath. exp. Neurol. *31:* 583–595 (1972).

3 Britt, R.; Lyons, B.; Pounds, D.; Prionas, S.: Biological feasibility of using ultrasound hyperthermia in the treatment of malignant brain tumors: tissue damage and thermal dosimetry studies. Proc. 10th Annu. Northeast Bioengineer. Conf. (IEEE Cat. No. 82: CHI 1747–5), pp. 101–104 (1982).

4 Britt, R.; Lyons, B.; Pounds, D.; Prionas, S.: Feasibility of using ultrasound hyperthermia in the treatment of malignant brain tumors: tissue damage threshold and thermal dosimetry studies. Med. Instrum. *17:* 172–177 (1983).

5 Douglas, M.A.; Parks, L.C.; Bebin, J.: Sudden myelopathy secondary to therapeutic total-body hyperthermia after spinal cord irradiation. New Engl. J. Med. *304:* 583–585 (1981).

6 Dubois, M.; Sato, S.; Lees, D.E.; Bull, J.M.; Smith, R.; White, B.G.; Moore, H.; MacNamara, T.E.: Electroencephalographic changes during whole body hyperthermia in humans. Electroenceph. clin. Neurophysiol *50:* 486–495 (1980).

7 Dubois, M.; Coppola, R.; Buchsbaum, M.S.; Lees, D.E.: Somatosensory evoked potentials during whole body hyperthermia in humans. Electroenceph. clin. Neurophysiol. *52:* 157–162 (1981).

8 Dunn, F.; Fry, F.J.: Ultrasonic threshold dosages for the mammalian central nervous system. IEEE Trans. biomed. Engng *18:* 253–256 (1971).

9 Gerweck, L. E.; Richards, B.: Influence of pH on the thermal sensitivity of cultured human glioblastoma cells. Cancer Res. *41:* 845–849 (1981).

10 Goffinet, D. R.; Choi, K. Y.; Brown, J. M.: The combined effects of hyperthermia and ionizing radiation on the adult mouse spinal cord. Radiat. Res. *72:* 238–245 (1977).

11 Hahn, G.: Hyperthermia and cancer, pp. 179–226 (Plenum Press, New York 1982).

12 Harris, A. B.; Erickson, L.; Kendig, J. H.; Mingrino, S.; Goldring, S.: Observations on selective brain heating in dogs. J. Neurosurg. *19:* 514–521 (1962).

13 Johnston, R. L.; Dunn, F.: Ultrasonic absorbed dose, dose rate, and produced lesion volume. Ultrasonics *14:* 153–155 (1976).

14 Kremkau, F. W.: Cancer therapy with ultrasound: a historical review. J. clin. Ultrasound *7:* 287–300 (1979).

15 Lele, P. P.: Hyperthermia by ultrasound; in Proc. Int. Symp. on Cancer Therapy by Hyperthermia and Radiation, pp. 168–178 (American College of Radiology, Washington 1975).

16 Lele, P. P.: Induction of deep, local hyperthermia by ultrasound and electromagnetic fields. Problems and choices. Radiat. environ. Biophys. *17:* 205–217 (1980).

17 Lele, P. P.: Thresholds and mechanisms of ultrasonic damage to 'organized' animal tissues; in Hazzard, Litz, Symposium on biological effects and characteristics of ultrasound sources; DHEW Publ. FDA 78-8048, pp. 224–239 (Government Printing Office, Washington 1977).

R. H. Britt, MD, PhD, Neurosurgery Department, Stanford University School of Medicine, Stanford, CA 94305 (USA)

Appendix: Supplementary Reference List

Outline

I. Oncogenes

Adamson, E.D.: High levels of expression of cellular oncogenes in extraembryonal tissues. Proc. Am. Ass. Cancer Res. *24:* 42 (1983).

Alitalo, K.; Schwab, M.; Lin, C.C., et al.: Homogeneously staining regions contain amplified copies of an abundantly expressed cellular oncogene (c-myc) in malignant neuroendocrine cells from a human colon carcinoma. Proc. natn. Acad. Sci. USA *80:* 1707–1711 (1983).

Schwab, M.; Alitalo, K.; Klempnauer, K.H., et al.: Amplified DNA with limited homology to the *myc* cellular oncogene is shared by human neuroblastoma cell lines and a neuroblastoma tumor. Nature, Lond. (in press).

Schwab, M.; Alitalo, K.; Varmus, H.E., et al.: A cellular oncogene (c-ki-*ras*) is amplified, overexpressed, and located within karyotypic abnormalities in mouse adrenocortical tumor cells. Nature, Lond. *303:* 497–501 (1983).

II. Tumor Growth

Dolbeare, F.; Gratzner, H.; Pallavicini, M., et al.: Flow cytometric measurement of total DNA content and incorporated bromodeoxyuridine. Proc. natn. Acad. Sci. USA (in press, 1983).

Kim, K.; Blechman, W.J.; Riddle, V.G., et al.: Basis of observed resistance of L1210 leukemia in mice: methotrexate, 6-thioguanine, 6-methylmercaptopurine, 6-mercaptopurine, 5-fluorouracil and 1-β-*D*-arabinofuranosylcytosine administered in different combinations. Cancer Res. *41:* 4529–4534 (1981).

Meyer, J.S.: Growth and cell kinetic measurements in human tumors. Pathol. A., Pt.2, *16:* 53–81 (1981).

Miller, B.E.; Miller, F.R.; Leith, J., et al.: Growth interaction in vivo between tumor subpopulations derived from a single mouse mammary tumor. Cancer Res. *40:* 3977–3981 (1980).

Molnar, P.; Groothuis, D.; Blasberg, R.G., et al.: Regional incorporation of 14-C-thymidine in tumor and brain tissue. J. Cerebral Blood Flow Metab. *3:* Suppl. 1, pp. S35–S37 (1983).

Nicolson, G.L.; Custead, S.E.: Tumor metastasis is not due to adaptation of cells to a new organ environment. Science *215:* 176–178 (1982).

Salcman, M.; Scott, E.; Schepp, R.S., et al.: A transplantable canine glioma model for use in experimental neuro-oncology. Neurosurgery *11:* 372–381 (1982).

Schiffer, D.; Giordana, M.T.; Soffietti, R., et al.: Histological observations on the regrowth of malignant gliomas after radiotherapy and chemotherapy. Acta neuropath. *58:* 291–299 (1982).

Steel, G.G.: Growth kinetics of tumours, p.351 (Clarendon Press, Oxford 1977).

Steel, G.G.: Growth kinetics of brain tumours; in Thomas, Graham, Brain tumours. Scientific basis, clinical investigations and current therapy, pp. 10–20 (Butterworths, London 1980).

III. Tumor Angiogenesis Factor

Brem, S.: The role of vascular proliferation in the growth of brain tumors. Clin. Neurosurg. *23:* 440–453 (1976).

Deane, B.R.; Lantos, P.L.: The vasculature of experimental brain tumours. I. A sequential light and electron microscope study of angiogenesis. J. neurol. Sci. *49:* 55–66 (1981).

Denekamp, J.; Hobson, B.: Endothelial-cell proliferation in experimental tumours. Br. J. Cancer *46:* 711–720 (1982).

Folkman, J.: Tumor angiogenesis: therapeutic implications. New Engl. J. Med. *285:* 1182–1186 (1971).

Folkman, J.; Haudenschild, C.: Angiogenesis in vitro. Nature, Lond. *288:* 551–556 (1980).

Folkman, J.; Langer, R.; Linhardt, R.J.; Haudenschild, C.; Taylor, S.: Angiogenesis inhibition and tumor regression caused by heparin or a heparin fragment in the presence of cortisone. Science *221:* 719–725 (1983).

Harris-Hooker, S.A.; Gajdusek, C.M.; Wight, T.N., et al.: Neovascular responses induced by cultured aortic endothelial cells. J. cell. Physiol. *114:* 302–310 (1983).

Kelly, P.J.; Suddith, R.L.; Hutchison, H.T., et al.: Endothelial growth factor present in tissue culture of CNS tumors. J. Neurosurg. *44:* 342–346 (1976).

IV. Selectivity of Treatment

A. Tumor Cell Sensitivity

1. Microcytotoxicity Assay

Kornblith, P.L.; Smith, B.H.; Leonard, L.A.: Response of cultured human brain tumors to nitrosoureas: correlation with clinical data. Cancer *47:* 255–265 (1981).

2. Precursor Uptake and Proliferation Assays

Ali-Osman, F.; Maurer, H.R.: Comparison of cytostatic sensitivities of L1210 cells and human stimulated lymphocytes in three cell proliferation assays. J. Cancer Res. clin. Oncol. *98:* 221–231 (1980).

Bogdahn, U.: Chemosensitivity of human malignant brain tumors: preliminary results. J. Neurooncol. *1:* 149–166 (1983).

Group for Sensitivity Testing of Tumors (KSST): In vitro short-term test to determine the resistance of human tumors to chemotherapy. Cancer *48:* 2127–2135 (1981).

Volm, M.; Wayss, K.; Kaufmann, M., et al.: Pretherapeutic detection of tumor resistance and the results of tumor chemotherapy. Eur. J. Cancer *15:* 983–993 (1979).

3. Multicellular Spheroid Studies

Darling, J.L.; Oktar, N.; Thomas, D.G.T.: Multicellular tumour spheroids derived from human brain tumours. Cell Biol. int. Rep. *7:* 23–30 (1983).

Deen, D.F.; Hoshino, T.; Williams, M.E., et al.: Development of a 9L multicellular spheroid system and its response to BCNU and radiation. J. natn. Cancer Inst. *64:* 1373–1382 (1980).

Pertuiset, B.F.; Rosenblum, M.L.; Poisson, M., et al.: Cinétique de sphéroïdes multicellulaires développés à partir d'un glioblastoma humain et action de la 1,3-bis (2-chloroéthyl)-1-nitrosouréa. Sem. Hôp. Paris *53:* 468–472 (1983).

Rathmer, K.; Weizsäcker, M.; Hamda, H., et al.: Methods of preclinical testing of malignant brain tumours. Proc. 33rd Annu. Meet. D. Ges. Neurochir., Kiel, 1982. Acta neurochir. *62:* 145–146 (1982).

4. Subrenal Capsule Assay

Bogden, A.E.; Cobb, W.R.; Lepage, D.J., et al.: Chemotherapy responsiveness of human tumors as first transplant generation xenografts in the normal mouse: six-day subrenal capsule assay. Cancer *48:* 10–20 (1981).

B. Normal Tissue Considerations

Gray, J.W.: Quantitative cytokinetics: cellular response to cell cycle phase specific agents. J. Pharmac. exp. Ther. (in press, 1983).

Gray, J.W.; Pallavicini, M.G.: Ara-C scheduling: theoretical and experimental considerations. Med. Pediat. Oncol., suppl. 1, pp. 93–108 (1982).

Pallavicini, M.G.; Gray, J.W.: Ara-C cytokinetic studies in normal tissues in vivo. Med. Pediat. Oncol., suppl. 1, pp. 109–123 (1982).

Tubiana, M.: Cell kinetics and radiation oncology. Int. J. Radiat. Oncol. Biol. Phys. *8:* 1471–1489 (1982).

C. Sensitizing and Protecting Agents

Blumberg, A.L.; Nelson, D.F.; Gramkowski, M., et al.: Clinical trials of WR-2721 in radiation therapy. Int. J. Radiat. Oncol. Biol. Phys. *8:* 561–564 (1982).

Brady, L.W.: Radiation sensitizers: their use in the clinical management of cancer. Cancer Management Series, vol. 5 (Masson, New York 1980).

EORTC Brain Tumour Group: Misonidazole in radiotherapy of supratentorial malignant brain gliomas in adult patients: a randomized double-blind study. Eur. J. Cancer clin. Oncol. *19:* 39–42 (1983).

Horowitz, M.; Blasberg, R.G.; Molnar, P., et al.: Regional brain tumor misonidazole distribution. Cancer Res. (in press).

Millar, J.L.; McElwain, T.J.; Clutterbuck, R.D., et al.: The modification of melphalan toxicity in tumor bearing mice by S-2-(3-amino-propylamino)-ethylphosphorothioc acid (WR-2721). Am. J. clin. Oncol. *5:* 312–328 (1982).

Philips, T.L.: Radiation sensitizers and protectors; in DeVita, Hellman, Rosenberg, Cancer, principles and practice of oncology, pp. 1822–1836 (Lippincott, Philadelphia 1982).

Sutherland, R.M.: Chemical modification: radiation and cytotoxic drugs. Special Issue Int. J. Radiat. Oncol. Biol. Phys. *8:* 323–809 (1982).

Tanaka, Y.; Sugahara, T.: Clinical experiences of chemical radiation protection in tumor radiotherapy in Japan; in Brady, Radiation sensitizers, pp. 421–425 (Masson, New York 1980).

Yuhas, J.M.: Chemical radiosensitization as a means of improving solid tumor radiotherapy; in Sokol, Maickel, Radiation-drug interactions in the treatment of cancer, p. 137 (Wiley, New York 1980).

Yuhas, J.M.: On the potential application of radioprotective drugs in solid tumor radiotherapy; in Sokol, Maickel, Radiation-drug interactions in the treatment of cancer, p. 113 (Wiley, New York 1980).

Yuhas, J.M.: Protective drugs in cancer therapy: optimal clinical testing and future directions. Int. J. Radiat. Oncol. Biol. Phys. *8:* 513–517 (1982).

V. CNS Toxicity from Therapy

A. Radiation Therapy

Burger, P.C.; Mahaley, M.S., Jr.; Dudka, L., et al.: The morphologic effects of radiation administered therapeutically for intracranial gliomas: a postmortem study of 25 cases. Cancer *44:* 1256–1272 (1979).

Fishman, M.L.; Bean, S.C.; Cogan, D.G.: Optic atrophy following prophylactic chemotherapy and cranial radiation for acute lymphocytic leukemia. Am. J. Ophthal. *82:* 571–576 (1976).

Harris, J.R.; Levene, M.B.: Visual complications following irradiation for pituitary adenomas and craniopharyngiomas. Radiology *120:* 167–171 (1976).

Henson, P.A.; Urich, H.: Cancer and the nervous system, pp. 573–621 (Blackwell Scientific, Oxford 1982).

Hochberg, F.H.; Slotnick, B.: Neuropsychologic impairment in astrocytoma survivors. Neurology, Minneap. *30:* 172–177 (1980).

Kinsella, T.J.; Weichselbaum, R.R.; Sheline, G.E.: Radiation injury of cranial and peripheral nerves; in Gilbert, Kagen, Radiation damage to the nervous system, pp. 145–153 (Raven Press, New York 1980).

Raimond, A.J.; Tomita, T.: The disadvantages of prophylactic whole CNS postoperative radiation therapy for medulloblastoma; in Paoletti, Walker, Butti, Knerich, Multidisciplinary aspects of brain tumor therapy, pp. 209–218 (Biomedical Press, Elsevier/North Holland, Amsterdam 1979).

Richards, G.E.: Effects of irradiation on the hypothalamic and pituitary regions; in Gilbert, Kagen, Radiation damage to the nervous system, pp. 175–180 (Raven Press, New York 1980).

Schatz, N.J.; Lichtenstein, S.; Corbett, J.J.: Delayed radiation necrosis of the optic nerves and chiasm. Neuroophthalmology *8:* 131–139 (1975).

Schiffer, D.; Giordana, M.T.; Paoletti, P., et al.: Pathology of human malignant gliomas after radiation and chemotherapy. Acta neurochir. *53:* 205–216 (1980).

Schiffer, D.; Giordana, M.T.; Paoletti, P., et al.: Radio- and chemotherapy of malignant gliomas. Pathological changes in the normal nervous tissue. Acta neurochir. *58:* 37–58 (1981).

B. Chemotherapy

Crafts D.C.; Levin, V.A.; Nielsen, S.: Intracarotid BCNU (NSC-409962): a toxicity study in six rhesus monkeys. Cancer Treatm. Rep. *60:* 541–545 (1976).

Egorin, M.J.; Bellis, E.H.; Salcman, M., et al.: The pharmacology of diazoquinone (NSC 182986) administered by intravenous intracarotid infusion to normal and intracranial tumor bearing puppies. J. Neurosurg. (submitted, 1983).

Greenberg, H.S.; Ensminger, W.D.; Seeger, J.F., et al.: Intra-arterial BCNU chemotherapy for the treatment of malignant gliomas of the central nervous system: a preliminary report. Cancer Treatm. Rep. *65:* 803–810 (1981).

Griffin, J.D.; Garnick, M.B.: Eye toxicity of cancer chemotherapy. A review of the literature. Cancer *48:* 1539–1549 (1981).

Vizel, M.; Oster, M.W.: Ocular side effects of cancer chemotherapy. Cancer *49:* 1999–2002 (1982).

VI. Photoradiation Therapy

Diamond, I.; McDonagh, A.F.; Wilson, C.B., et al.: Photodynamic therapy of malignant tumours. Lancet *ii:* 1175–1177 (1972).

Granelli, S.G.; Diamond, I.; McDonagh, A.F., et al.: Photochemotherapy of glioma cells by visible light and hematoporphyrin. Cancer Res. *35:* 2567–2570 (1975).

Laws, E.R., Jr.; Cortese, D.A.; Kinsey, J.H., et al.: Photoradiation therapy in the treatment of malignant brain tumors: a phase I (feasibility) study. Neurosurgery *9:* 672–678 (1981).

Little, F.M.; Gomer, C.J.; Hyman, S., et al.: Quantitative kinetics of tritium-labelled hematoporphyrin derivatives in normal and neoplastic tissues in the rat brain model (Abstract). AANS Meeting, Washington (1983).

Perria, C.; Capuzzo, T.; Cavagnaro, G., et al.: First attempts at the photodynamic treatment of human gliomas. J. Neurosurg. Sci. *24:* 119–129 (1980).

Rounds, D.E.; Jacques, S.; Shelden, C.H., et al.: Development of a protocol for photoradiation therapy of malignant brain tumors. Part I. Neurosurgery *11:* 500–505 (1982).

Wharen, R.E., Jr.; Anderson, R.E.; Laws, E.R.: Quantitation of hematoporphyrin derivative in human gliomas, experimental central nervous system tumors, and normal tissues. Neurosurgery *12:* 446–450 (1983).

VII. Chemotherapy Agents in Use

Aroney, R.S.; Kaplan, R.S.; Salcman, M., et al.: A phase II trial of aziridinylbenzoquinone (AZQ) (NSC 182986) in patients with recurrent primary brain tumors. Cancer Treatm. Rep. (in press, 1983).

Aronin, P.A.; Mahaley, J.S., Jr.; Rudnick, S.A., et al.: Prediction of BCNU pulmonary toxicity in patients with malignant gliomas: an assessment of risk factors. New Engl. J. Med. *303:* 183–188 (1980).

Edwards, M.S.; Levin, V.A.; Wilson, C.B.: Brain tumor chemotherapy: an evaluation of agents in current use for phase II and III trials. Cancer Treatm. Rep. *64:* 1179–1205 (1980).

EORTC Brain Tumour Group: Evaluation of CCNU, VM-26 plus CCNU and procarbazine in supratentorial brain gliomas. J. Neurosurg. *55:* 27–31 (1981).

Green, S.B.; Byar, D.P.; Walker, M.D., et al.: Comparisons of carmustine, procarbazine, and high-dose methylprednisolone as additions to surgery and radiotherapy for the treatment of malignant glioma. Cancer Clin. Trials (in press, 1983).

Levin, V.A.; Vestnys, P.A.; Wara, W.M., et al.: Improvement in survival produced by sequential therapies in the treatment of recurrent medulloblastoma. Cancer *51:* 1364–1370 (1983).

Levin, V.A.; Wilson, C.B.; Davis, R., et al.: A phase III comparison of BCNU, hydroxyurea, and radiation therapy to BCNU and radiation therapy for the treatment of primary malignant gliomas. J. Neurosurg. *51:* 526–232 (1979).

Schold, S.C.: Chemotherapy of primary central nervous system neoplasms. Semin. Neurol. *1:* 189–202 (1981).

Shapiro, W.R.: Treatment of neuroectodermal brain tumors. Ann. Neurol. *12:* 231–237 (1982).

Walker, M.D.; Alexander, E., Jr.; Hunt, W.E., et al.: Evaluation of BCNU and/or radiotherapy in the treatment of anaplastic gliomas: a cooperative clinical trial. J. Neurosurg. *49:* 333–343 (1978).

Walker, M.D.; Green, S.B.; Byar, D.P., et al.: Randomized comparisons of radiotherapy and nitrosoureas for the treatment of malignant glioma after surgery. New Engl. J. Med. *303:* 1323–1329 (1980).

Weir, B.; Band, P.; Urtasun, R., et al.: Radiotherapy and CCNU in the treatment of high-grade supratentorial astrocytomas. J. Neurosurg. *45:* 129–134 (1976).

VIII. Tumor Diagnostic Studies

A. CT Scans

Baker, H.L., Jr.; Houser, O.W.; Campbell, J.K.: National Cancer Institute study: evaluation of computed tomography in the diagnosis of intracranial neoplasms. Radiology *136:* 91–96 (1980).

Butler, A.R.; Horii, S.C.; Kricheff, I.I., et al.: Computed tomography in astrocytomas. A statistical analysis of the parameters of malignancy and the positive contrast-enhanced CT scan. Radiology *129:* 433–439 (1978).

Hayman, A.L.; Evans, R.A.; Hinck, V.C.: Delayed high dose contrast computed tomography. Cranial neoplasm. Radiology *136:* 677–684 (1980).

Hoffman, W.F.; Levin, V.A.; Wilson, C.B.: Evaluation of malignant glioma patients during the postirradiation period. J. Neurosurg. *50:* 624–628 (1979).

Latchaw, R.E.; Payne, J.T.; Loewenson, R.B.: Predicting brain tumor histology: change of effective atomic number with contrast enhancement. Am. J. Neurosurg. Res. *1:* 289–294 (1980).

Levin, V.A.; Hoffman, W.F.; Heilbron, D.C., et al.: Prognostic significance of the pretreatment CT scan on time to progression for patients with malignant gliomas. J. Neurosurg. *52:* 642–647 (1980).

McGeachie, R.E.; Gold, L.H.A.; Latchaw, R.E.: Periventricular spread of tumor demonstrated by computed tomography. Radiology *125:* 407–410 (1977).

Norman, D.; Stevens, E.A.; Wing, D., et al.: Quantitative aspects of contrast enhancement in cranial computed tomography. Radiology *129:* 683–688 (1978).

Salcman, M.; Levine, H.; Rao, K.: The value of sequential CT-scanning in the combined modality treatment of glioblastoma. Neurosurgery *8:* 15–19 (1981).

B. NMR Scans

Brant-Zawadski, M.; Davis, P.L.; Crooks, L.E., et al.: NMR demonstration of cerebral abnormalities: comparison with CT. Am. J. Neuroradiol. *4:* 117–124 (1983).

Buonanno, F.S.; Pykett, I.L.; Brady, T.J., et al.: Clinical relevance of two different nuclear magnetic resonance (NMR) approaches to imaging of a low grade astrocytoma. J. Comput. assist. Tomogr. *6:* 529–535 (1982).

Bydder, G.M.; Steiner, R.E.; Young, I.R., et al.: Clinical NMR imaging of the brain: 140 cases. Am. J. Neuroradiol. *3:* 459–480 (1982).

Crooks, L.E.; Mills, C.M.; Davis, P.L., et al.: Visualization of cerebral and vascular abnormalities by NMR imaging. The effects of imaging parameters on contrast. Radiology *144:* 843–852 (1982).

Gräfin von Einsiedel, H.; Löffler, W.: Nuclear magnetic resonance imaging of brain tumors unrevealed by CT. Eur. J. Radiol. *2:* 226–234 (1982).

Jezernik, M.; Sentjurc, M.; Schara, M.: Proton magnetic resonance characterization of brain tumors. Acta neurochir. *67:* 1–9 (1983).

C. Radionuclide Scans

Cowan, R.J.; Watson, N.E.: Special characteristics and potential of single photon emission computed tomography in the brain. Semin. nucl. Med. *10:* 333–344 (1980).

Ell, P.J.; Khan, O.: Emission computerized tomography: clinical applications. Semin. nucl. Med. *11:* 50–60 (1981).

Kuhl, D.E.; Barrio, J.R.; Huang, S.-C., et al.: Quantifying local cerebral blood flow by N-isopropyl-p-[^{123}I]iodoamphetamine (IMP) tomography. J. nucl. Med. *23:* 196–203 (1982).

Patronas, N.J.; Di Chiro, G.; Brooks, R.A., et al.: Work in progress: [^{18}F]-fluorodeoxyglucose and positron emission tomography in the evaluation of radiation necrosis of the brain. Radiology *144:* 885–889 (1982).

Winchell, H.S.; Horst, W.D.; Braun, L., et al.: N-isopropyl-p-I-123-iodoamphetamine: single pass brain uptake and washout; binding to brain synapsomes; and localization in dog and monkey brain. J. nucl. Med. *21:* 947–950 (1980).

Yen, C.; Yano, Y.; Budinger, T.F., et al.: Brain tumor evaluation using RB-82 and positron emission tomography. J. nucl. Med. *23:* 532–537 (1982).

IX. Surgery

A. Localization Techniques

1. Ultrasound

Chandler, W.F.; Knake, J.E.; McGillicuddy, J.E., et al.: Intraoperative use of real-time ultrasonography in neurosurgery. J. Neurosurg. *57:* 157–163 (1982).

Gooding, G.A.W.; Edwards, M.S.B.; Rabkin, A.E., et al.: Intraoperative real-time ultrasound in the localization of intracranial neoplasms. Radiology *146:* 459–462 (1983).

2. CT Stereotaxy

Appuzzo, M.L.; Sabshin, J.K.: Computed tomograpic guidance stereotaxis in the management of intracranial mass lesions. Neurosurgery *12:* 277–285 (1983).

Boethius, J.; Bergström, M.; Greitz, T.: Stereotaxic computerized tomography with a GE 8800 scanner. J. Neurosurg. *52:* 794–800 (1980).

Mackay, A.R.; Gutin, P.H.; Hosobuchi, Y., et al.: Computed tomography-directed stereotaxy for biopsy and interstitial irradiation of brain tumors: technical note. Neurosurgery *11:* 38–42 (1982).

Ostertag, C.B.; Mennel, H.D.; Kiessling, M.: Stereotactic biopsy of brain tumors. Surg. Neurol. *14:* 275–283 (1980).

B. Surgical Tools

1. Laser

Edwards, M.S.B.; Boggan, J.E.; Fuller, T.A.: The laser in neurological surgery: a review. J. Neurosurg. (in press, 1983).

Kelly, P.J.; Alker, G.J., Jr.; Goerss, S.: Computer-assisted stereotactic laser microsurgery for the treatment of intracranial neoplasms. Neurosurgery *10:* 324–331 (1982).

2. Ultrasound

Epstein, F.; Epstein, N.: Surgical treatment of spinal cord astrocytomas of childhood. A series of 19 patients. J. Neurosurg. *57:* 685–689 (1982).

X. Miscellaneous

Burger, P.C.; Dubois, P.J.; Schold, S.C., Jr., et al.: Computerized tomographic and pathologic studies of untreated, quiescent, and recurrent glioblastoma multiforme. J. Neurosurg. *58:* 159–169 (1983).

Rosenblum, M.L.; Gerosa, M.A.; Dougherty, D.V., et al.: Age-related chemosensitivity of stem cells from human malignant brain tumors. Lancet *i:* 885–887 (1982).

Salcman, M.: Survival in glioblastoma-historical perspective. Neurosurgery *7:* 435–439 (1980).

Salcman, M.; Kaplan, R.S.; Ducker, T.B., et al.: Effect of age and reoperation on survival in the combined modality treatment of malignant astrocytoma. Neurosurgery *10:* 454–463 (1982).

Salcman, M.; Kaplan, R.S.; Samaras, G.M., et al.: Aggressive multimodality therapy based on a multicompartmental model of glioblastoma. Surgery, St Louis *92:* 250–259 (1982).

Young, B.; Oldfield, E.H.; Markesbery, W.R., et al.: Reoperation for glioblastoma. J. Neurosurg. *55:* 917–921 (1981).

Cumulative Subject Index, vols. 27 and 28

Figures in italics refer to volume number.